Topics in Applied Physics Volume 43

Topics in Applied Physics Founded by Helmut K. V. Lotsch

Two-Dimensional Digital Signal Processing II

Transforms and Median Filters

Edited by T. S. Huang

With Contributions by
J.-O. Eklundh   T. S. Huang   B. I. Justusson
H. J. Nussbaumer   S. G. Tyan   S. Zohar

With 49 Figures

Springer-Verlag Berlin Heidelberg New York 1981

Professor *Thomas S. Huang*, Ph.D.
Department of Electrical Engineering and
Coordinated Science Laboratory, University of Illinois
Urbana, IL 61801, USA

ISBN 3-540-10359-7 Springer-Verlag Berlin Heidelberg New York
ISBN 0-387-10359-7 Springer-Verlag New York Heidelberg Berlin

Library of Congress Cataloging in Publication Data. Main entry under title: Two-dimensional digital signal processing II. (Topics in applied physics; v. 43) Bibliography: p. Includes index. 1. Image processing–Digital techniques. 2. Digital filters (Mathematics) 3. Transformations (Mathematics) I. Huang, Thomas S., 1936– TA1632.T9   621.38'0433   80-24775

© by Springer-Verlag Berlin Heidelberg 1981
Printed in Germany

Monophoto typesetting, offset printing and bookbinding: Brühlsche Universitätsdruckerei, Giessen
2153/3130-543210

Preface

Activities in digital image processing have been increasing rapidly in the past decade. This is not surprising when one realizes that in a broad sense image processing means the processing of multidimensional signals and that most signals in the real world are multidimensional. In fact, the one-dimensional signals we work with are often collapsed versions of multidimensional signals. For example, speech is often considered as a one-dimensional signal, viz., a function of a single variable (time). However, speech originally exists in space and therefore is a function of 4 variables (3 spatial variables and time).

There are analog (optical, electro-optical) as well as digital techniques for image processing. Because of the inherent advantages in digital techniques (flexibility, accuracy), and because of the rapid progress in computer and related technologies, such as LSI, and VLSI, it is fair to say that except for some specialized problems, digital techniques are usually preferred.

The purpose of this book and its companion (Two-Dimensional Digital Signal Processing I: Linear Filters) is to provide in-depth treatments of three of the most important classes of digital techniques for solving image processing problems: Linear filters, transforms, and median filtering. These two books are related but can be used independently.

In the earlier volume 6 of this series, *Picture Processing and Digital Filtering* (first edition, 1975), selected topics in two-dimensional digital signal processing including transforms, filter design, and image restoration, were treated in depth. Since then, a tremendous amount of progress has been made in these areas. In 1978 when we were planning on a second edition of that book (published in 1979), a decision was made not to make significant revisions but only to add a brief new chapter surveying the more recent results. And we projected that in-depth treatments of some of the important new results would appear in future volumes of the Springer physics program.

These two present books on two-dimensional digital signal processing represent the first two of these projected volumes. The material is divided into three parts. In the first part on linear filters, which is contained in the companion volume, major recent results in the design of two-dimensional nonrecursive and recursive filters, stability testing, and Kalman filtering (with applications to image enhancement and restoration) are presented. Among the highlights are the discussions on the design and stability testing of half-plane recursive filters, a topic of great current interest.

The second and third parts are contained in this volume. In the second part on transforms, two topics are discussed: algorithms for transposing large matrices, and number-theoretic techniques in transforms and convolution. Here we have a detailed derivation of the Winograd Fourier transform algorithm.

In the first and the second parts, the main concern is linear processing. In the third part on median filtering, a particular nonlinear processing technique is studied. Median filtering has become rather popular in image and speech processing. However, published results on it have been scarce. The two chapters of the third part contain new results most of which are published here for the first time.

The chapters in this volume are tutorial in nature, yet they bring the readers to the very forefront of current research. It will be useful as a reference book for working scientists and engineers, and as a supplementary textbook in regular or short courses on digital signal processing, image processing, and digital filtering.

Urbana, Illinois, September 1980   *Thomas S. Huang*

Contents

Contributors

Eklundh, Jan-Olof
   National Defense Research Institute (FOA), P.O. Box 1165
   S-581 11 Linköping, Sweden

Huang, Thomas S.
   Department of Electrical Engineering and
   Coordinated Science Laboratory, University of Illinois
   Urbana, IL 61801, USA

Justusson, Bo I.
   Department of Mathematics, Royal Institute of Technology
   S-100 44 Stockholm 70, Sweden

Nussbaumer, Henri J.
   IBM Centre d'Etudes et Recherches
   F-06610 LaGaude, France

Tyan, Shu-Gwei
   M/A-COM Laboratories, 11717 Exploration Lane
   Germantown, MD 20767, USA

Zohar, Shalhav
   Jet Propulsion Laboratory
   California Institute of Technology, 4800 Oak Grove Drive
   Pasadena, CA 91103, USA

1. Introduction

T. S. Huang

With 3 Figures

It is the goal of this volume to present in-depth treatments on two topics: two-dimensional digital transforms, and median filtering. The mathematical techniques to be discussed are motivated by applications to image processing.

There are three major areas of image processing [1.1]: efficient coding, restoration and enhancement, and pattern recognition. Many image restoration and enhancement techniques make use of linear spatially invariant (LSI) filters. Such filters are discussed in detail in [1.1, 2]. Transforms and related methods enable computationally efficient implementations of LSI filters. Many transforms are also useful in efficient coding and in feature extraction for pattern recognition.

Successful image restoration and enhancement often require nonlinear techniques. One such technique, which has become popular recently not only in image processing but also in signal processing in general, is median filtering. It can also be used in several pattern recognition-related tasks such as thinning and the extraction of small isolated objects in an image.

More detailed discussions on the chapters contained in the present book are given in the following sections.

1.1 Transforms

Various two-dimensional transforms were treated in a unified manner (based on outer products) in [Ref. 1.2, Chap. 2]. For a more detailed discussion of some of these transforms, see [1.3]. Among these transforms Fourier transform undoubtedly has the widest application. The others are mainly useful in image coding and occasionally in pattern recognition. Experience in the past few years has indicated that among the image-independent transforms (hence $K-L$ transform is not included), the discrete cosine transform (DCT) and the slant transform give the best performance in image coding. On the theoretical side, *Yemini* and *Pearl* [1.4] have shown that DCT is asymptotically optimal for all finite-order Markov signals. *Jain* [1.5] recently introduced a new family of unitary transforms whose members include many known transforms such as DCT.

Singular value decomposition (SVD) and its applications were discussed in some depth in [Ref. 1.2, Chaps. 1, 2]. Recently, *Sahasrabudhe* and *Vaidya* [1.6] related the behavior of the singular values to the autocorrelation function of an image. *Sahasrabudhe* and *Kulkarni* [1.7] developed a new method of combating noise in image restoration using SVD.

The conventional way of computing DCT is using FFT [1.8, 9]. However, a recent fast algorithm promises to speed up by a factor of 6 [1.10].

To calculate two-dimensional transforms, such as Fourier and Hadamard transforms, on a computer whose core memory cannot hold the entire image, auxiliary memory such as disks are needed and matrix transposition is usually used. An efficient algorithm for matrix transposition was suggested by *Eklundh* in 1972 [1.11]. More recently, he has developed two new algorithms and some results on optimum strategy [1.12]. See also *Ari* [1.13]. Alternatively, two-dimensional DFT and similar transforms can be calculated without matrix transposition [14–18].

An interesting recent research area is the use of number-theoretic techniques in signal processing. *Rader* was the first person who suggested the use of number-theoretic transforms (e.g., Fermat number transform) to do high-speed two-dimensional convolution [1.19–23]. An excellent introduction to this idea is the section written by *Rader* in [1.24]. For detailed treatments, see [1.25, 26].

By using these number-theoretic transforms, impressive computational savings can be obtained in cases where i) the one-dimensional sequences to be transformed are relatively short, ii) considerable accuracy is needed, and iii) multiplication is more costly than addition.

Winograd applied number-theoretic techniques to the calculation of the discrete Fourier transform (DFT) [1.27–30]. Compared to FFT, the number of multiplications is drastically reduced while the number of additions remains approximately the same. For example, for a sequence of 1024 samples, FFT requires 12,288 multiplications and 26,624 additions; while for a sequence of 1008 samples, the Winograd Fourier transform (WFT) requires 4212 multiplications and 25,224 additions. For extensions and programming considerations, see [1.31–34]. Hardware for performing WFT has been constructed and incorporated in a high-speed signal processor [1.35]. Quantization errors (due to roundoff and coefficient quantization) for WFT in the fixed-point case were studied by *Patterson* and *McClellan* [1.36], who found that in general WFT requires one or two more bits for data representation to give an error similar to that of FFT.

The material on transforms in the present volume is devoted to two areas: matrix transposition algorithms, and number-theoretic techniques. Chapter 2 discusses several efficient algorithms for transposing large matrices as well as the direct transform method of *Anderson* [1.14]. Chapter 3 describes fast algorithms for digital convolution and Fourier transformation based on polynomial transforms. In some of the algorithms, a combination of polynomial transforms and the Winograd algorithm was used. A detailed derivation of the Winograd Fourier transform algorithm is presented in Chap. 4.

1.2 Median Filters

Linear shift-invariant (LSI) filters are useful in image restoration and enhancement applications. For example, they can be used in realizing Wiener filters for reducing image noise. However, in order to reduce noise but keep the edges in the image sharp, one has to use linear shift-varying (LSV) or nonlinear filters. The limitations of LSI filters in image restoration were discussed in [Ref. 1.2, Chaps. 1, 5].

Many LSV and nonlinear procedures for image restoration were presented in [Ref. 1.2, Chaps. 1, 5]. In Chap. 5 of our companion volume on linear filters [1.1], LSV Kalman filters for noise reduction and image restoration were presented. In Chaps. 5 and 6 of the present volume, we concentrate on a particular nonlinear process: median filtering. Median filters have been found effective in reducing certain types of noise and periodic interference patterns without severely degrading the signal [1.37–39]. They have become very popular in image and speech processing.

Because theoretical analysis of the behavior of median filters is very difficult, published results have been almost nonexistent. The two chapters we have here in our book contain mainly new results not available before in the open literature. Chapter 5 deals with the statistical properties of median filters. In particular, various properties of the outputs of median filters are reported where the input is Gaussian noise or the sum of a step function and Gaussian noise.

Chapter 6 deals with the deterministic properties of median filters. Of particular interest are the results on the fixed points of median filters. A fixed point of a median filter is a sequence (in one dimension) or an array (in two dimensions) which is unchanged by the median filtering. In Chap. 6, *Tyan* shows that, in one dimension, the fixed points of median filters are "locally monotonic" sequences with the exception of certain periodic binary sequences. Recently, by restricting the sequences to be finite in length, *Gallagher* and *Wise* [1.40] were able to eliminate this exception.

In Chap. 6, an efficient two-dimensional median filtering algorithm based on histogram modification [1.39] is briefly described. Real-time hardware implementation of median filters using digital selection networks were discussed by *Eversole* et al. [1.41] and *Narendra* [1.42]. *Ataman* [1.43] presented a radix method for finding the median which is based on binary representation of the picture elements in the filter window. He compared the speed and hardware complexity of his method with those of the histogram algorithm and the digital selection network method. *Reeves* and *Rostampour* [1.44] considered the implementation of median filters on a binary array processor. *Wolfe* and *Mannos* [1.45] developed a technique of implementing median filters on video-rate pipeline processors.

Chapters 5 and 6 are largely theoretical. As a complement, we present some experimental results here. Figure 1.1 shows several examples of fixed points of

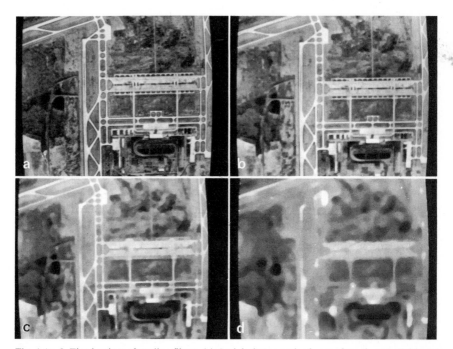

Fig. 1.1a–d. Fixed points of median filters. (a) Aerial photograph of part of an airport, digitized to 256 × 256 samples, 8 bits/sample. (b) Median filtered 6 times – 3 × 3 cross window. (c) Median filtered 6 times – 3 × 3 square window. (d) Median filtered 6 times – 5 × 5 square window

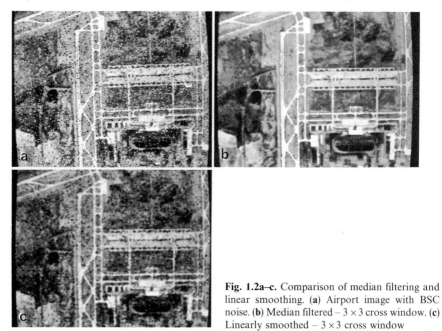

Fig. 1.2a–c. Comparison of median filtering and linear smoothing. (a) Airport image with BSC noise. (b) Median filtered – 3 × 3 cross window. (c) Linearly smoothed – 3 × 3 cross window

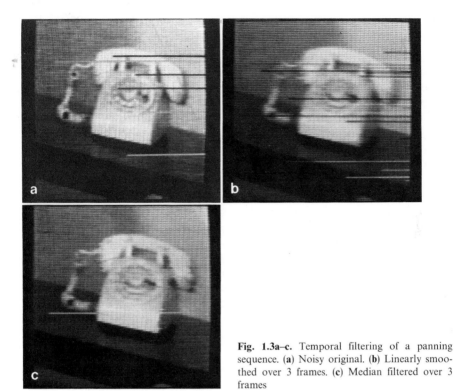

Fig. 1.3a–c. Temporal filtering of a panning sequence. (**a**) Noisy original. (**b**) Linearly smoothed over 3 frames. (**c**) Median filtered over 3 frames

median filters. In (a), we have the input to the median filters. In (b–d) we have the results of applying three different median filters, respectively, to (a) six times. Further application of the filters will not change the results appreciably. Therefore, the images in (b–d) are fixed points of the three median filters, respectively.

Median filters are especially suitable for reducing "salt and pepper" noise. This is illustrated in Fig. 1.2. In (a), we have the results of transmitting the image in Fig. 1.1a by PCM through a binary symmetrical channel. The resulting noise is of the "salt and pepper" type. The application of a median filter removes most of the noise – (b), while linear smoothing is not effective at all – (c).

Although, Chaps. 5 and 6 discuss two-dimensional (spatial) median filters, it is obvious that for moving images such as in television we can apply median filters in three dimensions (spatial and temporal), i.e., the filter window can be three dimensional. Temporal median filtering is especially suitable to reduce burst noise including line dropout. It also preserves motion much better than temporal averaging (linear smoothing). In [1.46], several temporal filtering experiments (including motion-compensated filtering) were described. In one experiment, a panning sequence containing white Gaussian noise and random line dropout was filtered temporally over three frames by a median filter and a linear smoother, respectively. The sequence was at 30 frames/s, each frame containing approximately 200 lines with 256 samples/line and 8 bits/sample. The panning was horizontal at a rate of approximately 5 picture elements per frame. Single frame results are shown in Fig. 1.3. In (a), we show a noisy original frame; (b), a linearly smoothed frame; and (c) a median filtered frame. It is to be noted that median filtering is much better in reducing line dropout and in preserving edge sharpness. However, linear smoothing is more effective in reducing Gaussian noise. These results are in accordance with the analysis in Chaps. 5 and 6.

Although median filtering, or for that matter linear smoothing, often improves the subjective quality of an image, it is by no means clear that they would facilitate further machine analysis of the image, such as pattern recognition or metric measurements. A comprehensive analysis of the effects of linear and median filtering on edge detection, shape analysis, and texture analysis is being carried out. Some results are given in [1.47].

References

1.1   T.S.Huang (ed.): *Two-Dimensional Digital Signal Processing I: Linear Filters*, Topics in Applied Physics, Vol. 42 (Springer, Berlin, Heidelberg, New York 1981)
1.2   T.S.Huang (ed.): *Picture Processing and Digital Filtering*. 2nd ed., (Springer, Berlin, Heidelberg, New York 1979)
1.3   N.Ahmed, K.R.Rao: *Orthogonal Transforms for Digital Signal Processing* (Springer, Berlin, Heidelberg, New York 1975)
1.4   Y.Yemini, J.Pearl: IEEE Trans. PAMI-1, 366–371 (1979)
1.5   A.K.Jain: IEEE Trans. PAMI-1, 356–365 (1979)
1.6   S.C.Sahasrabudhe, P.M.Vaidya: IEEE Trans. ASSP-27, 434–436 (1979)
1.7   S.C.Sahasrabudhe, A.D.Kulkarni: Comp. Graphics and Image Proc. 9, 203–212 (1979)
1.8   N.Ahmed, T.Natarjan, K.R.Rao: IEEE Trans. C-23, 90–93 (1974)
1.9   B.D.Tseng, W.C.Miller: IEEE Trans. C-27, 966–968 (1978)
1.10 W.H.Chen, C.H.Smith, S.C.Fralick: A fast computational algorithm for the discrete cosine transform, IEEE Trans. COM-25, 1004–1009 (1977)
1.11 J.O.Eklundh: IEEE Trans. C-21, 801 (1972)
1.12 J.O.Eklundh: Efficient matrix transposition with limited high-speed storage, FOA Reports, Vol. 12, No. 1, pp. 1–19, 1978, National Defense Research Institute, Linkoping, Sweden
1.13 M.B.Arl: IEEE Trans. C-27, 72–75 (1979)
1.14 G.L.Anderson: IEEE Trans. ASSP-28, 280–284 (1980)
1.15 M.Onoe: IEEE Proc. 63, 196–197 (1975)
1.16 I.DeLotto, D.Dotti: Comp. Graphics and Image Proc. 4, 271–278 (1975)
1.17 G.E.Rivard: IEEE Trans. ASSP-25, 250–252 (1977)
1.18 D.H.Harris, J.H.McClellan, D.S.K.Chan, H.W.Schuessler: Vector radix fast Fourier transform, 1977 IEEE Int. Conf. on ASSP Record, (1977) pp. 548–551
1.19 C.M.Rader: IEEE Trans. CS-22, 575 (1975)
1.20 R.C.Agarwal, C.S.Burrus: Proc. IEEE 63, 550–560 (1975)
1.21 I.S.Reed, T.K.Truong, Y.S.Kwoh, E.L.Hall: IEEE Trans. C-26, 874–881 (1977)
1.22 P.R.Chevillat: IEEE Trans. ASSP-26, 284–290 (1978)

1.23 B. Rice: IEEE Trans. ASSP-**27**, 432–433 (1979)

1.24 L. R. Rabiner, B. Gold: *Theory and Application of Digital Signal Processing* (Prentice-Hall, Englewood Cliffs, NJ 1975) pp. 419–434

1.25 J. McClellan, C. M. Rader: *Number Theory in Digital Signal Processing* (Prentice-Hall, Englewood Cliffs, NJ 1979)

1.26 H. Nussbaumer: *Fast Algorithms for the Computation of Convolutions and DFTs*, Springer Series in Information Sciences, Vol. 2 (Springer, Berlin, Heidelberg, New York 1981)

1.27 H. F. Silverman: IEEE Trans. ASSP-**25**, 152–165 (1977)

1.28 H. F. Silverman: IEEE Trans. ASSP-**26**, 268 und 482 (1978)

1.29 B. D. Tseng, W. C. Miller: IEEE Trans. ASSP-**26**, 268–269 (1978)

1.30 S. Zohar: IEEE Trans. ASSP-**27**, 409–421 (1979)

1.31 T. W. Parsons: IEEE Trans. ASSP-**27**, 398–402 (1979)

1.32 I. S. Reed, T. K. Truong: IEEE Trans. C-**28**, 487–492 (1979)

1.33 H. Nawab, J. H. McClellan: IEEE Trans. ASSP-**27**, 394–398 (1979)

1.34 L. R. Morris: IEEE Trans. ASSP-**26**, 141–150 (1978)

1.35 A. Peled: A Low-Cost Image Processing Facility Employing a New Hardware Realization of High-Speed Signal Processors, in *Advances in Digital Image Processing*, ed. by P. Stucki (Plenum, New York 1979)

1.36 R. W. Patterson, J. H. McClellan: IEEE Trans. ASSP-**26**, 447–455 (1978)

1.37 W. K. Pratt: "Median Filtering"; semiannual report, Image Processing Inst., Unvi. of Southern California, Sept. (1975) pp. 116–123

1.38 B. R. Frieden: J. Opt. Soc. Am. **66**, 280–283 (1976)

1.39 T. S. Huang, G. J. Yang, G. Y. Tang: IEEE Trans. ASSP-**27**, 13–18 (1979). A shorter version appeared in Proc. 1978 IEEE Conf. on Pattern Recognition and Image Processing, May 31-June 2, 1978, pp. 128–131

1.40 N. C. Gallagher, G. L. Wise: IEEE Trans. ASSP (to be published)

1.41 W. L. Eversole, D. J. Mayer, F. B. Frazee, T. F. Cheek, Jr.: "Investigation of VLSI Technologies for Image Processing", Proc. Image Understanding Workshop, Pittsburgh, Penn., pp. 191–195, Nov. (1978)

1.42 P. M. Narendra: IEEE Trans. PAMI (to be published). A short version appeared in Proc. 1978 IEEE Conf. on Pattern Recognition and Image Processing, Chicago, ILL., pp. 137–141; May (1978)

1.43 E. Ataman, V. K. Aatre, K. M. Wong: IEEE Trans. ASSP-**28**, 415–421 (1980)

1.44 A. P. Reeves, A. Rostampour: IEEE Trans. PAMI (to be published)

1.45 G. J. Wolfe, J. L. Mannos: "A Fast Median Filter Implementation", Proc. SPIE Seminar on Image Processing, Sept. 1979, San Diego, Ca.

1.46 T. S. Huang: "Noise filtering of moving images", Proc. Workshop on Automatic Tracking, Redstone Arsenal, Ala. Nov. 19–20 (1979)

1.47 G. Yang, T. S. Huang: "Median Filters and Their Applications to Image Processing"; Tech. Rpt., School of Electrical Engineering, Purdue University (1980)

2. Efficient Matrix Transposition

J. O. Eklundh

In this chapter we shall describe a number of different algorithms for transposition of matrices that can be accessed row- or columnwise and discuss their efficiency in terms of required memory and input and output operations. In particular, we shall dwell on matrices whose dimensions are highly composite numbers, e.g., powers of two. In fact, optimal algorithms will be presented in the latter case. We shall also discuss how an arbitrary matrix can be handled, by embedding it into a larger one with desirable properties. Most of the algorithms described require that each row (or column) can be addressed randomly, as in a random access file on disk, but some attention will also be given to the case when then rows are accessible only in sequence, as in a tape file. The first case is, however, of special interest, since then some of the methods allow square matrices to be transposed in-place, which in turn implies that any 2-D separable transform of an externally stored square matrix can be computed in-place as well.

We shall also present the direct approach. A comparison between the two approaches will show that they are essentially equal in performance in the cases that are common in the applications. It will, however, also be observed that important differences exist.

2.1 Background

Integral transforms, for example the Fourier transform, are important computational tools in digital signal processing. In two dimensions their discrete forms are double sums computed by summing iteratively in the two coordinate directions. It is straightforward to compute these double sums when the entire matrix can be stored in high speed memory. A problem arises when this is impossible or inconvenient. In many applications the matrix is stored on a block storage device, e.g., a disk, where the smallest record that can be accessed is an entire row or column. In this environment the direct computation of the transform is expensive because of the time involved in accessing the externally stored matrix.

One way of avoiding this problem is to transpose the matrix after the first summation. Another method due to *Anderson* [2.1] can be derived directly from the definition of the fast Fourier Transform (FFT), the crucial point being that the FFT is built up by transforms of subarrays and that all these transforms are computed in place.

Both these approaches are based on the separability of the Fourier transform. The 2-D discrete Fourier transform (DFT) of a complex array, $x(m, n)$, $m = 0, 1, \ldots, M-1$, $n = 0, 1, \ldots, N-1$ may be defined as

$$X(k, l) = \sum_{m=0}^{M-1} \sum_{n=0}^{N-1} x(m, n) \, W_M^{km} W_N^{ln}, \quad k = 0, 1, \ldots, M-1, l = 0, 1, \ldots, N-1 \quad (2.1)$$

where $W_P = \exp(-2\pi j/P)$ is used for notational convenience. Separability means that the DFT kernel $W_M^{km} W_N^{ln}$ is a product of two functions depending on, respectively, m and n alone. This allows us to compute the transform in two steps, first computing

$$Y(m, l) = \sum_{n=0}^{N-1} x(m, n) \, W_N^{ln} \quad (2.2a)$$

and then

$$X(k, l) = \sum_{m=0}^{M-1} Y(m, l) \, W_M^{km} \quad (2.2b)$$

where the first sum is an N-point 1-D DFT over the *rows* of the matrix $[x(m, n)]$ and the second is an M-point 1-D DFT over the *columns* of the resulting matrix $[Y(m, l)]$. Consequently the 2-D transform can be computed by the use of a 1-D transform and a matrix transposition. This is of particular interest because there is a "fast" algorithm for the 1-D DFT and, moreover, because this algorithm can be implemented in hardware.

On the other hand, the fact that the sum in (2.2b) is also a DFT implies that one can use the iterative summation technique of the FFT mixed with input and output of the transformed rows to compute $[X(k, l)]$ directly.

The transposition approach clearly applies to any separable transform, whereas the direct approach works when, in addition, a "fast" algorithm exists as, e.g., in the case of the Hadamard transform. Both methods also generalize to higher dimensions. (See [2.2] on how the DFT can be computed by use of transpositions).

More important, perhaps, is that also 1-D transforms of large arrays can be broken up into transforms of smaller arrays and matrix transpositions. In the case of the DFT this can be done in the following way (according to [2.2]): The DFT of the array $x(m)$, $m = 0, 1, \ldots, M-1$ can be defined as

$$X(k) = \sum_{m=0}^{M-1} x(m) \, W_M^{km} \quad k = 0, 1, \ldots, M-1. \quad (2.3)$$

Now, if $M = M_0 M_1$, we may write

$$m = m_1 M_0 + m_0 \quad 0 \leq m_0 < M_0, 0 \leq m_1 < M_1 \quad (2.4)$$

and

$$k = k_0 M_1 + k_1 \qquad 0 \leq k_0 < M_0, 0 \leq k_1 < M_1 \tag{2.5}$$

and consider x and X as matrices letting

$$x(m) \Leftrightarrow y(m_0, m_1) \tag{2.6}$$

and

$$X(k) \Leftrightarrow Y(k_1, k_0). \tag{2.7}$$

Since $W_M^{M_0 M_1} = 1$ and $W_M^{M_1} = W_{M_0}$, we get

$$Y(k_1, k_0) = \sum_{m_0=0}^{M_0-1} \sum_{m_1=0}^{M_1-1} y(m_0, m_1) W_M^{(m_0 + m_1 M_0)(k_1 + k_0 M_1)}$$

$$= \sum_{m_0=0}^{M_0-1} \sum_{m_1=0}^{M_1-1} y(m_0, m_1) W_M^{(m_0 + m_1 M_0)k_1} W_{M_0}^{m_0 k_0}. \tag{2.8}$$

Computing the sum in two steps we have

$$Z(m_0, k_1) = \sum_{m_1=0}^{M_1-1} y(m_0, m_1) W_M^{(m_0 + m_1 M_0)k_1}$$

$$m_0 = 0, 1, ..., M_0 - 1, k_1 = 0, 1, ..., M_1 - 1, \tag{2.9}$$

and

$$Y(k_1, k_0) = \sum_{m_0=0}^{M_0-1} Z(m_0, k_1) W_{M_0}^{m_0 k_0}$$

$$k_0 = 0, 1, ..., M_0 - 1, k_1 = 0, 1, ..., M_1 - 1. \tag{2.10}$$

Here, given the row of y, we first compute the corresponding *row* of Z by multiplication of the appropriate phase factors and summing. Then, given the *columns* of Z, we compute its DFT, which is the corresponding column of Y. To do this, again, we need to transpose Z.

Other methods have also been proposed for computing one- and multidimensional transforms with limited high-speed storage, some requiring matrix transposition (see, e.g., [2.2, 3]) and some not (see, e.g., [2.4, 5]). The latter approach is generally less efficient.

2.2 Methods for Transposing Externally Stored Matrices

2.2.1 Definition of Performance Criteria

In the next few sections we shall describe a number of methods for transposition of matrices stored on random access devices so that each record contains one row (or column; in the sequel we shall, however, always assume that the matrices are stored in row major order). Generally, all the algorithms achieve their task by reading a limited number of data records into high-speed memory, rearranging the data there, forming new output records and writing these on an output file, which may or may not coincide with the input file. Their performance will be defined in terms of the required number of high-speed memory locations, RM, and the required number of input and output operations (i.e., the number of records that have to be read and written), IO. These measures do not account for all the computational costs involved, but the remaining costs may be neglected or considered invariant when different methods are applied to the same matrix. More specifically, there are two main types of costs that are not included.

The first one is caused by the rearrangement of data in high-speed storage to form the new output records. However, one can create the new records *at output* without actually moving any data in storage. For example, it is possible to use implicit loops in the output statements in *Fortran* (see Sect. 2.7). Consequently, the CPU time needed is small and, moreover, it is proportional to the number of elements in the matrix times the number of passes over the data, which already is captured in the number of input and output statements. It should be noted that although we think of the algorithms as being implemented in this manner, in their descriptions we shall, for simplicity, still talk about transposing matrices and rearranging arrays in high-speed storage.

Secondly, by only counting the number of records that are transferred, we disregard the fact that the record lengths may vary when one or the other method is applied. Hence, it would be more accurate to account also for the number of words that have to be transferred. This cost can, however, be neglected for two reasons. First, the time needed to access one record is normally much higher than the time needed to transfer one word. For a modern high-speed disk the ratio between these two times typically exceeds 1000. Furthermore, a close look at the different methods will show that the variations in record lengths that occur generally are small compared to that ratio.

2.2.2 A Simple Block-Transposition Method

A simple and straightforward technique for transposition of an $M \times N$ matrix is presented in [2.1]. The matrix is transposed by first splitting up the original records of length N into records of length K, where K is chosen such that KM elements can be stored in high-speed storage. Then for $i = 1, ..., N/K$ the records

$j=1$, $i+N/K$, ..., $i+(M-1)N/K$ are read into a $K \times M$ matrix in high-speed storage which is subsequently transposed. K consecutive columns of the desired transpose are then obtained and can be written out.

The number of read and write operations needed for this algorithm is proportional to MN/K. To give good performance, K and hence the required amount of high-speed storage should be large. In fact, the method matches other proposed methods only when almost the entire matrix can be stored in high-speed memory. A similar performance is achieved by another block-transposition technique suggested by *Hunt* [2.3].

2.2.3 Transposition Using Square Partitions

The second algorithm was first proposed by *Eklundh* [2.6]. It is based on successive transposition of square partitions and forms a natural generalization of the method given by *Eklundh* [2.7]. It is also similar to a method used by *Ramapriyan* [2.8].

Let A be an $M \times N$ matrix and suppose that $\bar{M} = m_1 \ldots m_p \geq M$, where $m_i > 1$. Set $m_0 = 1$, $M_0 = M$, $N_0 = N$ and define for $i = 0, 1, \ldots, p$

$$P_i = m_1 \cdot \ldots \cdot m_i, \tag{2.11}$$

$$M_i = \left\lceil \frac{M_{i-1}}{m_i} \right\rceil \tag{2.12}$$

and

$$N_i = \left\lceil \frac{N_{i-1}}{m_i} \right\rceil \tag{2.13}$$

where $\lceil \ \rceil$ is defined by

$$\lceil x \rceil = \begin{cases} x, & x \text{ integer} \\ [x] + 1 & \text{otherwise,} \end{cases} \tag{2.14}$$

$[x]$ denoting the integer part of x. Then the transpose of A can be constructed in p steps as follows.

Step 1: Form M_1 matrices of size $m_1 \times N$ from consecutive rows of A. Each such matrix can, possibly after padding with zeros [at most $m_1(m_1 - 1)$], be partitioned into N_1 matrices of size $m_1 \times m_1$. These square matrices are now transposed in place. We obtain a matrix $A^{(1)}$, whose elements are $1 \times P_1$ matrices and the dimension of which is is $P_1 M_1 \times N_1$.

Inductively we define the ith step, when the first $i-1$ steps are completed.

Step i: In step *i* we start with a $P_{i-1}M_{i-1} \times N_{i-1}$ matrix, $A^{(i-1)}$, whose elements are $1 \times P_{i-1}$ matrices. Then M_i submatrices of size $m_i \times N_{i-1}$ are formed by selecting rows that are P_{i-1} rows apart in the following way: If λP_i rows, $\lambda = 0, 1, ..., M_i - 1$, have been processed we pick out m_i rows, P_{i-1} rows apart, starting with each of the ensuing P_{i-1} rows, that is starting with row $\lambda P_i + \mu$, $\mu = 1, 2, ..., P_{i-1}$. Hence, a typical submatrix consists of the rows (in order) $\lambda P_i + \mu + \nu P_{i-1}$, where $\nu = 0, 1, ..., m_i - 1$ and λ and μ are fixed in the ranges given above. Such a matrix can, if necessary after zero-filling, be partitioned into N_i matrices of size $m_i \times m_i$, whose elements are $1 \times P_{i-1}$ matrices. These square matrices are transposed, and a $P_i M_i \times N_i$ matrix $A^{(i)}$, with $1 \times P_i$ matrices as elements, is obtained. The $1 \times P_i$ matrices are parts of the rows of $\tilde{A}$, the transpose of A.

This description of the algorithm gives an intuitive feeling of how the rows of $\tilde{A}$ are constructed as parts of the records of $A^{(p)}$ and will also be useful in the analysis of the performance. However, it is not immediate that each element will end up in the right place. Therefore, we shall describe the algorithm in a different way without referring to partitions and so be able to prove that the algorithm works, at the same time that we indicate how it can be implemented.

First we look at the case when $M \geq N$. Let $[k^{(i)}, l^{(i)}]$ denote the position in $A^{(i)}$ (no longer considered as partitioned) of the element $a(k, l)$ of A, $i = 1, ..., p$, $k = 0, 1, ..., M - 1$, $l = 0, 1, ..., N - 1$. We then observe that k and l can be uniquely represented in the form

$$k = k_{p-1}P_{p-1} + ... + k_1 P_1 + k_0$$
$$l = l_{p-1}P_{p-1} + ... + l_1 P_1 + l_0 \tag{2.15}$$

where $0 \leq k_i < m_{i+1}$, $0 \leq l_i < m_{i+1}$ and P_i is given by (2.11). The representations (2.15) are evidently obtained from the principal remainders at successive divisions by $m_1, m_2, ..., m_p$, as in an ordinary number system representation. Let us simply use the notation

$$k = n(k_{p-1}, ..., k_1, k_0)$$
$$l = n(l_{p-1}, ..., l_1, l_0). \tag{2.16}$$

Now step one implies that A is partitioned into $m_1 \times m_1$ matrices which are transposed. This means that

$$k^{(1)} = n(k_{p-1}, ..., k_1, l_0)$$
$$l^{(1)} = n(l_{p-1}, ..., l_1, k_0) \tag{2.17}$$

since $k_1, ..., k_{p-1}$ and $l_1, ..., l_{p-1}$ just determine which $m_1 \times m_1$ matrix the element belongs to.

Inductively we assume that after $i-1$ steps we have

$$k^{(i-1)}=n(k_{p-1}, ..., k_i, k_{i-1}, l_{i-2}, ..., l_0)$$
$$l^{(i-1)}=n(l_{p-1}, ..., l_i, l_{i-1}, k_{i-2}, ..., k_0). \qquad (2.18)$$

Step i can then be described in the following way. Consider the matrix as partitioned into $P_i \times P_i$ matrices. Then, inside each such matrix, form $m_i \times m_i$ matrices by picking out m_i rows and the same m_i columns that are P_{i-1} rows (columns) apart in all possible ways. This means that $k_i, ..., k_{p-1}$ and $l_i, ..., l_{p-1}$ remain unchanged, since they determine the specific $P_i \times P_i$ matrix, and so do $l_{i-2}, ..., l_0$ and $k_{i-2}, ..., k_0$ because they determine which of the $m_i \times m_i$ matrices the element belongs to. In this particular matrix $a(k, l)$ occurs in row k_{i-1} and column l_{i-1} before transposition and so

$$k^{(i)}=n(k_{p-1}, ..., k_i, l_{i-1}, l_{i-2}, ..., l_0)$$
$$l^{(i)}=n(l_{p-1}, ..., l_i, k_{i-1}, k_{i-2}, ..., k_0). \qquad (2.19)$$

It follows that $[k^{(p)}, l^{(p)}]=(l, k)$, proving that $a(k, l)$ has the right position in $A^{(p)}$. We observe that the transposition is performed by a sequence of swaps over the digits in the mixed radix representation (2.15). In particular, if $m_i=2$ for all i, we get the algorithms described in [2.7, 9].

It only remains to establish the relation between $A^{(p)}$ and $\tilde{A}$. If $\bar{M}=M$, the first N rows of $A^{(p)}$ obviously are the rows of A, whereas the last $\bar{M}-N$ rows contain only zeros. In fact, they need never be created. If, on the other hand, $\bar{M}>M$, then the rows of $A^{(p)}$ also contain $\bar{M}-M$ zeros at the end, which like the zero rows, never have to be written out.

The case $M<N$ can be treated in a similar way, if in (2.15) we write

$$l=l_p P_p+l_{p-1}P_{p-1}+...+l_0 \qquad (2.20)$$

with $0 \le l_i < m_{i+1}$ for $i=0, 1, ..., p-1$ and $0<l_p$. The final position of $a(k, l)$ will now be $[k^{(p)}, l^{(p)}]$, where

$$k^{(p)}=l_{p-1}P_{p-1}+...+l_0 \qquad (2.21)$$

$$l^{(p)}=l_p P_p+k_{p-1}P_{p-1}+...+k_0. \qquad (2.22)$$

In this case the rows of $A^{(p)}$ contain one ($\bar{M} \ge N$) or several ($\bar{M}<N$) rows of $\tilde{A}$. Writing $A^{(p)}$ as partitioned into N_p $\bar{M} \times \bar{M}$ matrices, where N_p is defined by (2.13),

$$A^{(p)}=(B_1|B_2|...|B_{N_p}), \qquad (2.23)$$

we obtain $\tilde{A}$ as

$$
\begin{pmatrix} B_1 \\ \overline{B_2} \\ \vdots \\ B_{N_p} \end{pmatrix} = \left(\begin{array}{c|c} \tilde{A} & 0 \\ \hline 0 & 0 \end{array} \right). \tag{2.24}
$$

It should be observed that this operation as well as the cancelling of the zeros is performed at output in step p and without additional data handling.

If $M \neq N$ then, at least for some products $\bar{M}$, the algorithm will first transpose $K \times K$ matrices, where $K = \min(M, N)$ and then concatenate $(M < N)$ or split up $(M > N)$ the rows.

If the rows of $A^{(i)}$ are created at output, then the performance in terms of required high-speed storage locations is measured at step i by

$$
\text{RM}_\text{S} = \max_{1 \leq i \leq p} \{ m_i N_{i-1} P_{i-1} \}. \tag{2.25}
$$

M extra storage locations would allow us to actually form the rows of A in high-speed storage, which is of interest when 2-D transforms are computed. This can, however, also easily be done by transpositions in main storage using no extra storage if $m_p \leq N_p$ and using $m_p N_p$ additional locations if $m_p > N_p$ (see Sect. 2.7). It will turn out that for optimal embeddings generally $m_p N_p \ll M$.

One can note that performing the transpositions in main storage at each step, instead of just writing out the words in the right order, not only is slower but also requires more memory. More precisely, one gets

$$
\text{RM}_{\text{S}'} = \max_{1 \leq i \leq p} \{ m_i N_i P_i \}. \tag{2.26}
$$

We shall later (Sect. 2.3.1) show that $N_i P_i$ increases with i, which implies that $\text{RM}_{\text{S}'} \geq \text{RM}_\text{S}$.

If the zero rows of $A^{(p)}$ are not written out, the expression for the required number of input and output operations will be

$$
\text{IO}_\text{S} = M + 2 \sum_{i=1}^{p-1} M_i P_i + N. \tag{2.27}
$$

If $\bar{M} = M$, the algorithm presented here has the property that all the matrices $A = A^{(0)}, A^{(1)}, ..., A^{(p-1)}$ contain M rows, while $A^{(p)}$, if the zeros are discarded, contains N rows. This implies that the number of input and output operations depends only on the number of factors of M. If we, as mentioned earlier, disregard the varying (increasing) row lengths, we can optimize the algorithm for a given M and p by minimizing the required number of high-speed storage locations. It will be shown how a minimizing factorization of M can be characterized under certain conditions. If these conditions do not hold, the optimum can still be determined easily in any particular case. Moreover the dependence of the solution on p will be demonstrated. If $\bar{M} > M$, a factorization minimizing the memory requirements will no longer minimize the required number of input and output operations. However, it will be shown that the variation in number of IO operations between different factorizations with the same number of factors is fairly small. This suggests a simple algorithm, which iterates over increasing $\bar{M}$ and quickly finds the factorization minimizing the memory requirements for each p. Because $p \leq \lceil \log_2 M \rceil$ it is then easy to determine which of these solutions is cheapest in terms of memory *and* input/output operations. In general one can in this way determine a strategy which is close to the global optimum. The approach will be outlined in Sect. 2.4.

2.2.4 Floyd's Algorithm

The square partition algorithm has the property that, if all the factors of $\bar{M}$ are equal to 2 or powers of 2, then it can be implemented using only shift and masking operations. This follows from the description in Sect. 2.2.3 (see also [2.7]). Another algorithm, presented by *Floyd* [2.10], also has this property. It coincides with the square partition algorithm if $M = N = 2^m$ and it can be applied to any square matrix directly (without zero-filling).

An $M \times M$ matrix A can be transposed in $p = \log_2 M$ passes over data using $2M$ main storage locations in the following way:

1) Form $A^{(1)}$, where $a^{(1)}(k, l) = a(-k, l)$.
2) Form $B^{(0)}$, where $b^{(0)}(k, k+l) = a^{(1)}(k, l)$.
3) Recursively, for $i = 1, \ldots, p$, form $B^{(i)}$ from $B^{(i-1)}$ by shifting the columns for which the ith binary digit $[l_{i-1}$ in (2.15)] equals 1 2^i steps.
4) Form $\tilde{A}$ from $B^{(p)}$, setting $\tilde{a}(k, l) = b^{(p)}(k, k+l)$.

Observe that neither $A^{(1)}$ nor $B^{(0)}$ or $B^{(p)}$ actually have to be created. A proof of the algorithm is given in [2.10].

Using, say, $2^m M$ high-speed locations, one can execute m steps of the algorithm in one pass, as in the square partition algorithm. However, this algorithm then requires some additional bookkeeping, since the cycling of the elements spans over *all* the rows and therefore cannot be completed inside a given slice of 2^m rows.

Summarizing, we see that when

$$RM_F = 2^m M \tag{2.28}$$

high-speed locations are used, the number of IO operations is

$$IO_F = 2M \lceil \log_2 M/m \rceil . \tag{2.29}$$

In particular, it is shown in [2.10] that if M is a power of 2, then the right-hand side of (2.29) is also a lower bound on the number of IO operations needed to transpose an $M \times M$ matrix with the given memory. Hence, this algorithm as well as the square partition algorithm and the algorithm presented in the next section all minimize IO for given RM. It can be observed that the same lower bound on IO was derived by *Stone* [2.11], who also derived an algorithm attaining that bound. However, his algorithm is not directly applicable to this problem, neither can it be implemented using only shift and masking operations.

Equation (2.29) gives only an upper bound on the number of IO operations required. The optimal algorithm may in fact do strictly better. For instance, the optimal algorithm will for a 5×5 matrix with $m=1$ give IO$=28$ compared to the upper bound that is 30 [2.10]. No algorithm that gives minimal IO for given RM for *all* square matrices has, however, yet been presented. As it is described here, Floyd's algorithm will in general require more IO operations than the other algorithms we describe. However, these generally use somewhat more high-speed storage or external storage. An example of this will be given in Sect. 2.5.

Floyd's method can obviously also be generalized to rectangular matrices using the approach described in Sect. 2.2.3.

2.2.5 Row-in/Column-out Transposition

This algorithm is a generalization of an algorithm for transposing matrices stored on sequential devices proposed by *Knuth* [2.12].

The original algorithm transposes an $M \times M$ matrix stored sequentially (one record being one row) in $\lceil \log_2 M \rceil$ passes over the data, using $2^{\lceil \log_2 M \rceil + 1}$ high-speed storage locations and four additional sequential files as follows.

At each step two files are created which at the next step are processed synchronously:

Step 1:
File 1: $a_{11}a_{21}a_{12}a_{22} \cdots a_{1,M/2}a_{2,M/2} \cdots a_{1M}a_{2M}a_{51}a_{61}a_{52}a_{62} \cdots$
$\qquad a_{M-3,M}a_{M-2,M}$
File 2: $a_{31}a_{41}a_{32}a_{42} \cdots a_{3,M/2}a_{4,M/2} \cdots a_{3M}a_{4M}a_{71}a_{81}a_{72}a_{82} \cdots$
$\qquad a_{M-1,M}a_{M,M}$
Step 2:
File 3: $a_{11}a_{21}a_{31}a_{41}a_{12}a_{22}a_{32}a_{42} \cdots a_{1,M/4}a_{2,M/4}a_{3,M/4}a_{4,M/4} \cdots$
$\qquad a_{4,M}a_{91} \cdots a_{M-5,M}$
File 4: $a_{51}a_{61}a_{71}a_{81}a_{52}a_{62}a_{72}a_{82} \cdots a_{5,M/4}a_{6,M/4}a_{7,M/4}a_{8,M/4} \cdots$
$\qquad a_{8,M}a_{13,1} \cdots a_{M,M}$
$\vdots$

by reading one record from each input file and writing out the data in the new order as two records on one of the output files, letting these alternate. The transpose will obviously be obtained after $\lceil\log_2 M\rceil$ passes or after $M\lceil\log_2 M\rceil$ read and write operations.

Some slight modifications of the record sizes are necessary. In fact, we must increase (some of) the records so that at step i, $i=1,\ldots,\lceil\log_2 M\rceil-1$, they contain $k2^i$ elements where k is an even number. This means that (some of) the input records at the last step contain $2^{\lceil\log_2 M\rceil}$ elements.

A closer look at this algorithm shows that the ith step actually corresponds to a sequence of perfect shuffles (see [2.11]) of pairs of arrays, whose elements are 2^{i-1} element subarrays. In this sense Knuth's algorithm is indeed similar to the one proposed by *Stone* [2.11], but the shuffling is performed at output and the rearranged arrays of step i are not created in high-speed storage during that step. This suggests a slight modification of the algorithm, in which the last step transposes 2×2 matrices with elements that are $2^{\lceil\log_2 M\rceil-1}\times1$ matrices. In this way we obtain the rows of the transpose in high-speed storage *before* they are written out, which is useful if the algorithm is applied in connection with, e.g., the FFT.

This algorithm can evidently be generalized to rectangular matrices using the method given in Sect. 2.2.3. This is an efficient method for transposing sequentially stored matrices. For instance, it is generally more efficient than the approach suggested by *Schumann* [2.14, 15] (see [2.6] for a proof).

Let us now instead suppose that A is stored on a random access device. The algorithm then has a generalization that works exactly like the square partition algorithm except in the following respect. Given $A^{(i-1)}$, still built up out of $1\times P_{i-1}$ partitions that are parts of the rows of $\tilde{A}$, the rows of $A^{(i)}$ are formed by writing out m_i consecutive columns of the submatrices in high-speed storage. The rows of $A^{(i)}$ will then consist of parts of the rows of $\tilde{A}$ of size P_i. Usually there is no need to append any dummy data at output.

Note that, as can be seen in the binary case, this method and the square partition method are not identical. However, they demand the same amount of resources, that is

$$RM_{RC} = RM_S = \max_{1\leq i\leq p}\{m_i N_{i-1} P_{i-1}\} \tag{2.30}$$

and

$$IO_{RC} = IO_S = M + 2\sum_{i=1}^{p-1} M_i P_i + N. \tag{2.31}$$

They differ, though, in some respects that sometimes may be important. In-place swapping is no longer as simple as transposition of square matrices. But generally we are not concerned about this until the last step, and then the two methods will again work in the same way (see the examples in Sect. 2.5). A more important point can be made about the requirements on external storage. Applied to square matrices stored as random access files, both algorithms can work in-place on one single file. This is also true for the rectangular partition method presented below, which can then be forced to work with squares ($m_i = n_i$). On nonsquare matrices the three methods are no longer alike.

The original input file can always be a sequential file that is left unchanged. Aside from that, the square method can be applied in-place until step p, whereupon $\tilde{A}$ is stored in a second random access area. The rectangular method needs two random access areas to host $A^{(2i)}$ and $A^{(2i+1)}$, $i = 0, 1, \ldots$. These files will moreover vary in size. The row-in/column-out method, finally, can be implemented like the square method. If, however, at the last step, the order in which the groups of rows of $A^{(p-1)}$ are processed is altered (an example is given in Sect. 2.6), the rows of $\tilde{A}$ can be generated sequentially. Furthermore, the method works on sequential files if $\max_{1 \le i \le p} \{m_i\}$ auxiliary files are available.

There is, of course, also a rectangular version of the algorithm presented in this section. We leave the details to the reader.

2.2.6 The Rectangular Partition Algorithm

The last algorithm we shall present was originally proposed by *Delcaro* and *Sicuranza* [2.13]. Their algorithm is built up of successive transpositions of partitioned matrices of order $m_i \times n_i$, where m_i and n_i are factors of M and N, respectively. Later, *Ramapriyan* [2.8] generalized the algorithm to the case in which $\bar{M} = m_1 \cdot \ldots \cdot m_p \ge M$ and $\bar{N} = n_1 \cdot \ldots \cdot n_p \ge N$, where p, $m_1, \ldots, m_p$ and $n_1, \ldots, n_p$ are determined by numerical optimization of the required computation time.

The algorithm works as follows. Let A be an $M \times N$ matrix and suppose that $\bar{M} = m_1 \cdot \ldots \cdot m_p \ge M$ and $\bar{N} = n_1 \cdot \ldots \cdot n_p \ge N$, where p is any integer, $p \ge 2$. Set $m_0 = 1$, $n_0 = 1$, $M_0 = M$, $N_0 = N$ and define the following arrays of integers for $i = 0, 1, \ldots, p$.

$$P_i = m_0 \cdot m_1 \cdot \ldots \cdot m_i$$

$$Q_i = n_0 \cdot n_1 \cdot \ldots \cdot n_i$$

$$M_i = \lceil M_{i-1}/m_i \rceil \tag{2.32}$$

$$N_i = \lceil N_{i-1}/n_i \rceil.$$

Then at step i, $i = 1, 2, \ldots, p$ the input matrix $A^{(i-1)}$, where $A^{(0)} = A$, is considered as an $M_{i-1} \times N_{i-1}$ matrix, whose elements are $Q_{i-1} \times P_{i-1}$ matrices which in fact are submatrices of $\tilde{A}$. This partitioned matrix is embedded in an $M_i \times N_i$ matrix from which $m_i \times n_i$ matrices are selectively read into high-speed storage, transposed and written out on $A^{(i)}$. The rows are chosen as in [2.13] that is Q_{i-1} rows apart. $A^{(i)}$ will then be built up by $Q_i \times P_i$ blocks and at step p an $\bar{N} \times \bar{M}$ matrix containing $\tilde{A}$ is generated.

Alternatively one can look upon $A^{(i-1)}$ as being built up by $1 \times P_{i-1}$ partitions, since the $Q_{i-1} \times P_{i-1}$ are never actually formed. The rows of $A^{(i)}$ will then contain $1 \times P_i$ partitions, which form parts of the rows of $\tilde{A}$.

If $\bar{M} = M$ and $\bar{N} = N$ the algorithm coincides with the algorithm of [2.13]. Otherwise, the matrix $A^{(i-1)}$ is at step i embedded in the minimal matrix for which the algorithm [2.13] works, given $m_1, ..., m_p$ and $n_1, ..., n_p$.

The required memory can be written as

$$RM_R = \max_{1 \leq i \leq p} \{m_i N_{i-1} P_{i-1} + P_i N_i\}, \tag{2.33}$$

where the second term could be dropped except for $i = p$.

The number of input and output operations are

$$IO_R = M + 2 \sum_{i=1}^{p-1} Q_i M_i + N. \tag{2.34}$$

2.3 Optimization of the Performance of the Algorithms

2.3.1 Two Lemmas

All the algorithms presented in the previous sections, except Floyd's binary oriented algorithm, involve an embedding of the original matrix in a larger matrix. We shall now discuss how these embeddings should be chosen in order to optimize the performance of the algorithms. First we need two crucial lemmas.

Let $\bar{M}$ and N be arbitrary integers, $\bar{M} > 1$, $N > 1$, and assume that for some $p > 1$ $\bar{M} = m_1 \cdot ... \cdot m_p$, where $m_i > 1$, $i = 1, ..., p$. Define as before $m_0 = 1$, $p_i = m_0 \cdot ... \cdot m_i$, $i = 0, 1, ..., p$ and $N_0 = N$, $N_i = \lceil N_{i-1}/m_i \rceil$, $i = 1, ..., p$. We then have

Lemma 1. The array $(N_i P_i)_{i=0}^p$ is increasing.

Proof. By definition $N_i = \lceil N_{i-1}/m_i \rceil \geq N_{i-1}/m_i$, that is $N_i m_i \geq N_{i-1}$. Hence, multiplication by P_{i-1} gives $N_i P_i \geq N_{i-1} P_{i-1}$. Q.E.D.

This lemma simply states that the length of the rows increases in the square algorithm.

Lemma 2. $N/P_i \leq N_i < N/P_i + 1$, $i = 1, ..., p$.

Proof. We use induction on the number of factors of P_i. We have

$$N_1 = \begin{cases} N/m_1 & \text{if } m_1 \text{ is a divisor of } N \\ [N/m_1] + 1 & \text{otherwise}. \end{cases}$$

In the first case the proposition is immediate, in the second we only need to note that $[N/m_1] < N/m_1$.

Assume then that the inequality is true for $i-1$ factors, that is

$$\frac{N}{P_{i-1}} \leq N_{i-1} < \frac{N}{P_{i-1}} + 1.$$

(2.35)

Write $N_{i-1} = q_i m_i + r_i$ where $0 \leq r_i < m_i$. Then $q_i = [N_{i-1}/m_i]$ and

$$N_i = \begin{cases} q_i & \text{if } r_i = 0 \\ q_i + 1 & \text{if } 0 < r_i < m_i. \end{cases}$$

(2.36)

Using the left-hand side of (2.35) we find

$$N_i \geq \frac{N_{i-1}}{m_i} \geq \left(\frac{N}{P_{i-1}}\right) \Big/ m_i = \frac{N}{P_i}.$$

(2.37)

On the other hand, if $r_i = 0$, then

$$N_i = \frac{N_{i-1}}{m_i} < \left(\frac{N}{P_{i-1}} + 1\right) \Big/ m_i = \frac{N}{P_i} + \frac{1}{m_i} < \frac{N}{P_i} + 1.$$

(2.38)

Furthermore, if $1 \leq r_i < m_i$ then we may write

$$N_i = q_i + 1 = \frac{N_{i-1}}{m_i} + \frac{m_i - r_i}{m_i} < \frac{N_{i-1}/P_{i-1} + 1}{m_i} + \frac{m_i - r_i}{m_i}$$

$$= \frac{N}{P_i} + \frac{m_i - (r_i - 1)}{m_i} \leq \frac{N}{P_i} + 1. \quad \text{Q.E.D.}$$

(2.39)

We immediately get the following corollary:

Corollary. Since N_i is an integer, N_i and $N_i P_i$ are invariant under re-factorization of $P_i = m_1 \cdot \ldots \cdot m_i$. In particular $m_1, \ldots, m_i$ can be permuted.

Let us now introduce some new and more general notation. Denote finite arrays of positive integers by $[\alpha]$, $[\beta]$, ... or by enumeration $(m_1, m_2, \ldots, m_p)$. For a given array $[\alpha] = (m_1, m_2, \ldots, m_p)$, write its subarrays $[\alpha_j] = (m_1, \ldots, m_j)$ and set $N_{[\alpha_0]} = N, \ldots, N_{[\alpha_j]} = \lceil N_{[\alpha_{j-1}]}/m_j \rceil$ and $P_{[\alpha_0]} = 1, P_{[\alpha_j]} = m_1 \cdot \ldots \cdot m_j, j = 1, \ldots, p$.

2.3.2 The Square Partition Algorithm

Set $R_{[\alpha_j]} = m_j N_{[\alpha_{j-1}]} P_{[\alpha_{j-1}]}$ and $R[\alpha] = \max_{1 \leq j \leq p} \{R_{[\alpha_j]}\}$. We then have

Theorem 1. Let $\bar{M}$ and N be given integers, $\bar{M} > 1$, $N > 1$, and suppose that $\bar{M} = m_1 \cdot \ldots \cdot m_p$, where $m_1 \geq m_2 \geq \ldots \geq m_p > 1$. Let $[\alpha] = (m_1, \ldots, m_p)$ and assume that $\pi[\alpha]$ is an arbitrary permutation of $[\alpha]$. Then $R[\alpha] \leq R[\pi[\alpha]]$.

Let us before we prove the theorem make some observations. A look back at (2.25, 27) reveals that the theorem shows that the minimum memory requirements for the square partition methods, given the factorization, is obtained if the factors are ordered in a decreasing sequence. It is then assumed that the rows of $\tilde{A}$ are never formed (as consecutive arrays) in high-speed storage. But this requires at most $m_p N_p$ additional storage locations at step p (see [2.6]), and we shall see that the decreasing order will give the minimum then too. The theorem is also true if we consider $R'_{[\alpha_j]} = m_j N_j P_{[\alpha_j]}$, which is obtained if at each step we form the rows; in fact, a similar proof can be given.

The following examples show that there indeed is an order dependence.

Examples. If $N = 22$ and $\bar{M} = 5 \cdot 3 = 15$ we have $R_{(5)} = 110$ and $R_{(5, 3)} = 75$, while $R_{(3)} = 66$ and $R_{(3, 5)} = 120$, that is $R(3, 5) = 120 > 110 = R(5, 3)$. Setting $\bar{M} = 6 \cdot 5 = 30$ we see that $R(6, 5) = 132$ while $R(5, 6) = 150$, that is, this property does not depend on whether $\bar{M} < N$ or not. In these two examples, the maximum is attained for the maximal factor. This is not always so if $\bar{M} > N$, e.g., if $N = 22$ and $\bar{M} = 7 \cdot 6$ then $R_{(7)} = 154$ and $R_{(7, 6)} = 168$. If $\bar{M} \leq N$, however, R_{max} will always be attained for the last occurring maximum factor (see below). This proposition is however not true for the array $m_j N_j P_j, j = 1, ..., p$, obtained when swapping is done at each step.

Proof. Suppose that m_i is a factor for which the maximum $R[\alpha]$ is attained. Let $\pi[\alpha]$ be an arbitrary permutation of $[\alpha]$ and consider the shortest subarray at the beginning of $\pi[\alpha]$ containing all the elements $m_1, ..., m_i$. Denote this array by $[\bar{\beta}]$. If $[\bar{\beta}]$ has j elements, then $i \leq j \leq p$, and its last element is m_k, where $1 \leq k \leq i$. Let $[\beta]$ be the subarray obtained from $[\bar{\beta}]$ by deletion of m_k. Now reorder $[\bar{\beta}]$ to $[\bar{\gamma}]$ as follows. If $k = i$, let $[\bar{\gamma}] = [\bar{\beta}]$, but if $k < i$ $[\bar{\gamma}]$ is obtained from $[\bar{\beta}]$ by changing places between m_i and m_k. As before let $[\gamma]$ be the $j - 1$-element subarray at the beginning of $[\bar{\gamma}]$.

We now observe that since $P_{[\gamma]} \geq P_{[\beta]}$ it follows from Lemma 2 and its corollary that $N_{[\gamma]} \leq N_{[\beta]}$. (Note that Lemma 2 limits N_i to an interval containing exactly one integer). Moreover, $m_i P_{[\gamma]} = m_k P_{[\beta]}$, so that $m_k P_{[\beta]} N_{[\beta]} \geq m_i P_{[\gamma]} N_{[\gamma]}$.

Let $[\delta]$ be an array obtained from $[\gamma]$ by putting $m_1, ..., m_{i-1}$ at the beginning (in order). Then, again using the corollary of Lemma 2, we get $P_{[\delta]} N_{[\delta]} = P_{[\gamma]} N_{[\gamma]}$. Applying Lemma 1 to $[\delta]$ we then have

$$R[\pi[\alpha]] \geq m_k P_{[\beta]} N_{[\beta]} \geq m_i P_{[\gamma]} N_{[\gamma]} = m_i P_{[\delta]} N_{[\delta]} \geq m_i P_{i-1} N_{i-1} = R[\alpha].$$

Q.E.D.

This theorem does not account for the extra storage locations needed to form the rows of $\tilde{A}$ at step p. This can be done with $m_p N_p$ additional words at step p (see [2.6]). We then have

Corollary. Theorem 1 is true also for the array $(T_i)_{i=1}^p$,

$$T_i = m_i P_{i-1} N_{i-1}, i = 1, ..., p-1; T_p = m_p P_{p-1} N_{p-1} + m_p N_p.$$

Proof. Let $m'_1, \ldots, m'_p$ be a reordering of $m_1, \ldots, m_p$ and denote the corresponding array to be maximized by $(T'_i)^p_{i=1}$. Suppose that $\max\limits_{1 \leq i \leq p} T'_i = T'_j$. We then observe that $m'_p \geq m_p$, according to the assumptions, so that $P'_{p-1} \leq P_{p-1}$ and hence $N'_{p-1} \geq N_{p-1}$. That is, since $P'_p = P_p = \bar{M}$ and $N'_p = N_p$ we get $T'_p = P'_p N'_{p-1} + m'_p N'_p \geq P_p N_{p-1} + m_p N_p = T_p$. This proves the corollary, since

$$\max_{1 \leq i \leq p-1} T_i \leq \max_{1 \leq i \leq p} T'_i \text{ according to the theorem}. \quad \text{Q.E.D.}$$

Next we prove the nonsurprising fact that the best factorization into $p+1$ factors requires less high-speed storage than any factorization into p factors. Define for a given $\bar{M}$

$$\Omega_p(\bar{M}) = \{[\alpha] = (m_1, \ldots, m_p); m_i > 1, m_1 \cdot \ldots \cdot m_p = \bar{M}\} \tag{2.40}$$

and set

$$S_p(\bar{M}) = \min\{R_{[\alpha]}; [\alpha] \in \Omega_p(\bar{M})\}. \tag{2.41}$$

Theorem 2. If $\bar{M}$ can be factored into $p+1$ nontrivial factors then $S_{p+1}(\bar{M}) \leq S_p(\bar{M})$.

Proof. Let $[\alpha] = (m_1, \ldots, m_p)$ be an optimal factorization of $\bar{M}$ giving $S_p(\bar{M})$. At least one factor, say m_j, can then according to the assumptions be writen $m_j = st$, where s and t are integers $s > 1$, $t > 1$. Set $[\beta] = (m_1, \ldots, m_{j-1}, s, t, m_{j+1}, \ldots, m_p)$, $[\sigma] = (m_1, \ldots, m_{j-1}, s)$ and $[\tau] = (m_1, \ldots, m_{j-1}, s, t)$. From the corollary to Lemma 2 we know that $N_{[\tau]} = N_j$. Moreover, the same lemma shows that $N_{[\sigma]} \leq N_{j-1}$. Since $sP_{j-1} < P_j$ we then get

$$S_{p+1}(\bar{M}) \leq R_{[\beta]} = \max\{P_1 N_0, \ldots, P_{j-1} N_{j-2}, sP_{j-1} N_{j-1}, P_j N_{[\sigma]},$$

$$\cdot P_{j+1} N_j, \ldots, P_p N_{p-1}\}$$

$$\leq \max\{P_1 N_0, \ldots, P_{j-1} N_{j-2}, P_j N_{j-1}, P_{j+1} N_j, \ldots, P_p N_p^{-1}\} = S_p(\bar{M}). \tag{2.42}$$

$$\text{Q.E.D.}$$

This theorem is still true if the memory requirements at step i are set to $m_i P_i N_i$. The proof is then somewhat more complicated.

What is missing now is a theorem characterizing the optimal p-factor solution. One result in this direction is given by the following theorem.

Let us introduce

$$r_p(\bar{M}) = \min\{\|[\alpha]\|_\infty; [\alpha] \in \Omega_p(\bar{M})\}, \tag{2.43}$$

where $\|(m_1, \ldots, m_p)\|_\infty = \max\limits_{1 \leq i \leq p}\{m_i\}$. We then have

Theorem 3. Suppose that $\bar{M} \leq N$. Then $R[\alpha] = S_p(\bar{M})$ only if $\|[\alpha]\|_\infty = r_p(\bar{M})$. In fact, if there are several elements in $\Omega_p(\bar{M})$ having minimum norm, then those with the fewest occurrences of this value, $r_p(\bar{M})$, will minimize $S_p(\bar{M})$.

Proof. According to Theorem 1, we only have to consider decreasing arrays. Let $[\alpha] = (m_1, ..., m_p) \in \Omega_p$. Then, since $\bar{M} \leq N$, Lemma 2 gives

$$m_i P_{i-1} N_{i-1} < P_i(N/P_{i-1} + 1) = m_i N + P_i \leq (m_i + 1)N. \tag{2.44}$$

This implies that $m_i = m_1$ must hold for the maximum. If $[\alpha]$ and $[\alpha']$ are two decreasing arrays in $\Omega_p(\bar{M})$ such that $r_p(\bar{M}) = m_1 < m_1'$ then evidently

$$m_i P_{i-1} N_{i-1} < (m_1 + 1)N \leq m_1' N \leq R[\alpha'] \tag{2.45}$$

so that $R[\alpha] < R[\alpha']$. The latter part of the theorem can be seen from (2.25) and Lemma 1, if the decreasing order is used.   Q.E.D.

This theorem implies that the memory requirements (for fixed p) are minimum when the sizes of the factors vary as little as possible. It is, however, not true if $\bar{M} > N$. If, for instance, $\bar{M} = 9 \cdot 9 \cdot 4 \cdot 4 = 6 \cdot 6 \cdot 6 \cdot 6$ and $N = 243$ then $R(9, 9, 4, 4) = 2187$, whereas $R(6, 6, 6, 6) = 2592$. Nonetheless, the theorem can be used to simplify the search for the optimal embedding, as will be shown in Sect. 2.4. It can be noted that the theorem does not hold if one considers l^1-norms or the array $m_i P_i N_i$ (see [2.6]). One can also observe that even if $M < N$, the best factorization, $\bar{M}$, could satisfy $\bar{M} > N$.

Unfortunately, the required number of input and output operations [see (2.27)] will not be minimized if the factors are ordered in decreasing sequence. If, e.g., $M = 52$ then $\bar{M} = 5 \cdot 4 \cdot 3$, $3 \cdot 4 \cdot 5$, and $4 \cdot 5 \cdot 3$ give $IO_s = 52 + 230 + N$, $52 + 228 + N$, and $52 + 224 + N$, respectively. If $\bar{M} = M$ then $IO_s = (2p + 1)M + N$ independently of the order of the factors. Generally, we know from Lemma 2 that $\bar{M} \leq M_i P_i \leq \bar{M} + P_i - 1$, that is

$$M + N + 2p\bar{M} \leq IO_s^{(p)} \leq M + N + 2p\bar{M} + 2\sum_{i=1}^{p-1}(P_i - 1). \tag{2.46}$$

The required number of additional input/output operations caused by increasing M to $\bar{M}$ is, since $m_i > 1$, thus not greater than

$$2\sum_{i=1}^{p-1}(P_i - 1) = 2\left[\frac{\bar{M}}{m_p}\left(\frac{1}{m_2 \cdot ... \cdot m_{p-1}} + ... + \frac{1}{m_{p-1}} + 1\right)\right.$$
$$\left. - (p-1)\right] < 2\left[2\frac{\bar{M}}{m_p} - (p-1)\right]. \tag{2.47}$$

This corresponds to less than $2/m_p$ of a data pass.

Using the fact that some rows contain dummy elements from step 2 on, one can of course decrease this number further. Such a procedure will, however, complicate the implementation of the algorithm.

A special case worth mentioning is the one in which $\bar{M}$ and N can be written $\bar{M} = m_1 \cdot ... \cdot m_p$, $N = m_1 \cdot ... \cdot m_{p-1} \cdot k$. This ordering will then result in exactly

$2(p-1)\bar{M}+N$ input/output operations, independently of the magnitude of the factors. If $k\geq m_p$, that is if $N\geq M$, no extra storage is needed at step p, even if the rows of $\tilde{A}$ are formed. From Theorem 3 it then follows that this ordering will require an optimal $\max\limits_{1\leq i\leq p} m_i N$ storage locations. At the same time the number of input/output operations needed will in general be less than what is obtained using the decreasing order. If $k<m_p$ the latter statement is still true, but the memory requirements will increase slightly.

Examples. Let $M=60$ and $N=72$. Then factoring $M=5\cdot4\cdot3=3\cdot4\cdot5$ gives in both cases a required memory of 360, while the numbers of input/output operations are 442 and 372, respectively. If, on the other hand, $N=24$, we get $RM=126$ and 130, while $IO=214$ and 180, respectively.

2.3.3 The Rectangular Partition Algorithm

We are now ready to look at the rectangular partition algorithm. This means that from now on $\bar{N}=n_1\cdot\ldots\cdot n_p$, $N_0=N$ and $N_i=\lceil N_{i-1}/n_i\rceil$. Disregarding the additional storage locations needed to form the rows at each step (also step p) we have

Theorem 4. If $(m_i)_{i=1}^p$ increases and $(n_i)_{i=1}^p$ decreases then $(P_iN_{i-1})_{i=1}^p$ cannot have a strict maximum in the interior. In particular, this ordering will minimize the memory requirements and satisfy

$$\max_{1\leq i\leq p} \{P_iN_{i-1}\}\leq\max\left\{\min_{1\leq i\leq p} \{m_iN\},\ \min_{1\leq i\leq p} \{n_i+1\}\,\bar{M}\right\}.$$

Proof. Assume that a strict maximum is obtained for P_iN_{i-1}, where $1<i<p$. Then $P_{i+1}N_i<P_iN_{i-1}$ so that according to the definition of N_i

$$0<P_iN_{i-1}-P_{i+1}N_i=P_i(N_{i-1}-m_{i+1}N_i)\leq P_i\left(N_{i-1}-m_{i+1}\frac{N_{i-1}}{n_i}\right)$$

$$=P_iN_{i-1}\left(1-\frac{m_{i+1}}{n_i}\right), \tag{2.48}$$

that is $m_{i+1}<n_i$. The assumptions give $m_i\leq m_{i+1}<n_i\leq n_{i-1}$, implying $m_i+1\leq n_{i-1}$ (since we are considering integers).

We also have $P_iN_{i-1}>P_{i-1}N_{i-2}$. From Lemma 2 we know $N_{i-1}<N_{i-2}/n_{i-1}+1$, that is $(N_{i-1}-1)n_{i-1}<N_{i-2}$. This means that

$$0<P_iN_{i-1}-P_{i-1}N_{i-2}=P_{i-1}(m_iN_{i-1}-N_{i-2})$$

$$<P_{i-1}\{m_iN_{i-1}-(N_{i-1}-1)n_{i-1}\}\leq P_{i-1}\{m_iN_{i-1}-(N_{i-1}-1)(m_i+1)\}$$

$$=P_{i-1}(m_i-N_{i-1}+1). \tag{2.49}$$

Since we deal with integers we conclude that $N_{i-1} \leq m_i \leq n_i$. But then $N_i = 1$, so that $P_{i+1} N_i = P_{i+1} = m_{i+1} P_i$, and hence since $N_{i-1} \leq m_i$ we have $P_i N_{i-1} \leq P_i m_i \leq P_i m_{i+1} = P_{i+1} N_i$, contradicting the assumption.

We observe that if $P_{i+1} N_i < P_i N_{i-1}$, Lemma 2 will yield strict inequality in (2.49) even if $P_i N_{i-1} = P_{i-1} N_{i-2}$. This means that the array can have no interior plateaus, because then $P_{i+1} N_i < P_i N_{i-1}$ must hold for some i, $1 < i < p$.

Obviously, we always have $\max_{1 \leq i \leq p} \{P_i N_{i-1}\} \geq \max\{(m_1 N, \bar{M} N_{p-1})\}$, with equality if the ordering proposed is used. Moreover, this bound is minimal if m_1 and N_{p-1} are minimal. But we have earlier noted that N_{p-1} is a minimum if Q_{p-1} is a maximum, that is if $n_p = \min_{1 \leq i \leq p} \{n_i\}$. Finally we infer from Lemma 2 that

$$N_{p-1} < N/Q_{p-1} + 1 \leq \bar{N}/Q_{p-1} + 1 = n_p + 1. \quad \text{Q.E.D.}$$

Examples. The following examples show that the array can have a minimum in the interior. Moveover, an interior maximum may occur if the order is reversed. Consequently there is no converse theorem giving an upper bound to the memory requirements.

(m_1, m_2, m_3)	(n_1, n_2, n_3)	N	$(P_1 N, P_2 N_1, P_3 N_2)$
(2, 4, 6)	(4, 3, 2)	19	(38, 40, 96), increasing
(2, 3, 5)	(6, 4, 2)	47	(94, 48, 60), interior min.
(2, 3, 4)	(8, 5, 3)	115	(230, 90, 72), decreasing
(10, 9, 8)	(2, 3, 4)	19	(190, 900, 2160), increasing
(6, 4, 2)	(2, 3, 4)	19	(114, 240, 144), interior max.
(2, 3, 4)	(4, 5, 6)	119	(238, 180, 144), decreasing.

To create the rows of $\tilde{A}$ at step p, $m_p N_{p-1}$ additional storage locations are needed if $m_p > N_{p-1}$ (see [2.6]).

The orderings of Theorem 4 will then no longer guarantee a minimum. If, for instance, $N = 19$, $(n_1, n_2, n_3) = (4, 3, 2)$, then the required number of storage locations is $96 + 12 = 108$ if $(m_1, m_2, m_3) = (2, 4, 6)$ but $96 + 8 = 104$ if $(m_1, m_2, m_3) = (2, 6, 4)$. From Theorem 4, since $N_{p-1} \leq n_p$, we can conclude that, for given $\bar{M}$ and N,

$$\min\{\text{RM}_R\} \leq \max\{\underline{m}N, (\underline{n}+1)\bar{M} + \bar{m}\underline{n}\} \tag{2.50}$$

where

$$\underline{m} = \min_{1 \leq i \leq p} \{m_i\}, \quad \bar{m} = \max_{1 \leq i \leq p} \{m_i\} \text{ and } \underline{n} = \min_{1 \leq i \leq p} \{n_i\}.$$

An important property of the rectangular partition method is that there is a direct trade-off between the number of memory allocations and the input and output required. This is seen from the following theorem.

Theorem 5. The sum $M + 2 \sum_{i=1}^{p-1} M_i Q_i + N$ is a minimum if $(m_i)_{i=1}^p$ is decreasing and $(n_i)_{i=1}^p$ increasing, and a maximum if $(m_i)_{i=1}^p$ is increasing and $(n_i)_{i=1}^p$ decreasing.

Proof. We prove the first part by induction on p. Let $p=2$ and $m_1 \geq m_2$, $n_1 \leq n_2$. Introduce $M_1^{(1)} = \lceil M/m_1 \rceil$ and $M_1^{(2)} = \lceil M/m_2 \rceil$. Evidently $M_1^{(1)} \leq M_1^{(2)}$ so we get $M_1^{(1)} n_1 \leq M_1^{(1)} n_2$, $M_1^{(1)} n_1 \leq M_1^{(2)} n_1 \leq M_1^{(2)} n_2$, proving the assertion for $p=2$.

Now suppose the assertion is true for $p-1$ factors and consider arbitrary factorizations $\bar{M} = m_1 \cdot \ldots \cdot m_p$, and $\bar{N} = n_1 \cdot \ldots \cdot n_p$. First reorder $m_1, \ldots, m_{p-1}$ to $m_1' \geq \ldots \geq m_{p-1}'$ and $n_1, \ldots, n_{p-1}$ to $n_1' \leq \ldots \leq n_{p-1}'$ and put $m_p' = m_p$ and $n_p' = n_p$. Then, according to Lemma 2, $M_{p-1}' = M_{p-1}$. Obviously $Q_{p-1}' = Q_{p-1}$, so that the induction assumption gives $\sum_{i=1}^{p-1} M_i' Q_i' \leq \sum_{i=1}^{p-1} M_i Q_i$.

If $m_p' \leq m_{p-1}'$ the m_i's are now decreasing. Otherwise set $m_p'' = m_{p-1}'$ and $m_{p-1}'' = m_p'$ and $m_i'' = m_i'$, $i = 1, \ldots, p-2$. This gives $M_{p-1}'' \leq M_{p-1}'$ while $M_i'' = M_i'$, $i = 1, \ldots, p-2$, so $\sum_{i=1}^{p-1} M_i'' Q_i' \leq \sum_{i=1}^{p-1} M_i' Q_i'$. Finally, if $(m_i'')_{i=1}^{p-1}$ is not decreasing we can again apply the theorem for $(p-1)$ factors to obtain $m_1''' \geq \ldots \geq m_{p-1}'''$, $m_p''' \geq m_p''$, where $\sum_{i=1}^{p-1} M_i''' Q_i' \leq \sum_{i=1}^{p-1} M_i'' Q_i'$.

Correspondingly, if $n_p' < n_{p-1}'$ set $n_i'' = n_i'$, $i = 1, \ldots, p-2$, $n_{p-1}'' = n_p'$ and $n_p'' = n_{p-1}'$. Then $Q_i'' = Q_i'$, $i = 1, \ldots, p-2$ and $Q_{p-1}'' < Q_{p-1}'$, so $\sum_{i=1}^{p-1} M_i''' Q_i'' < \sum_{i=1}^{p-1} M_i''' Q_i'$. Using the assertion for $(p-1)$ factors once more we get $n_1''' \leq \ldots \leq n_{p-1}'''$ and $n_p''' = n_p''$, where $\sum_{i=1}^{p-1} M_i''' Q_i''' \leq \sum_{i=1}^{p-1} M_i''' Q_i'' \leq \sum_{i=1}^{p-1} M_i Q_i$ and $(m_i''')_{i=1}^p$ is decreasing and $(n_i''')_{i=1}^p$ is increasing.

The second part of the theorem follows from a similar argument. Q.E.D.

2.3.4 On the Advantages of Inserting the Factor 1

In the theorems above concerning the square partition algorithm, we have assumed that the factors $m_1, \ldots, m_p$ are all greater than 1. This assumption is in fact not necessary. It is easy to see that if any unit factor is deleted, the required memory will be the same, while the number of input/output operations will decrease.

The rectangular partition algorithm does not possess this property unless the unit factor occurs in the interior of the array. We have more precisely

Proposition 2. Let $\bar{M} = m_1 \cdot \ldots \cdot m_p$, and $\bar{N} = n_1 \cdot \ldots \cdot n_p$. Then, if $m_k = 1$, for some k, $1 < k \leq p$, there exist factorizations $m_1' \cdot \ldots \cdot m_{p-1}' = \bar{M}$ and $n_1' \cdot \ldots \cdot n_{p-1}' = \bar{N}$ for which

$$\max_{1 \leq i \leq p-1} \{P_i' N_{i-1}'\} \leq \max_{1 \leq i \leq p} \{P_i N_{i-1}\} \quad \text{and} \quad \sum_{i=1}^{p-2} M_i' Q_i' \leq \sum_{i=1}^{p-1} M_i Q_i.$$

The same assertion can be made if $n_k = 1$ and $1 \leq k < p$.

Proof. If $m_k = 1$ the new factorizations are given by $m'_i = m_i$, $i = 1, ..., k-1$, $m'_i = m_{i+1}$, $i = k, ..., p-1$, $n'_i = n_i$, $i = 1, ..., k-2$, $n'_{k-1} = n_k n_{k-1}$, $n'_i = n_{i+1}$, $i = k, ..., p-1$. Because $N'_{k-1} = N_k$ the memory requirements will then be $\max_{\substack{1 \leq i \leq p \\ i \neq k}} \{p_i N_{i-1}\}$. In the sum for the number of input/output operations the term $M_k Q_k$ will be deleted.

If $n_k = 1$ we set $m'_k = m_k m_{k+1}$ and keep the other factors to get the same result.   Q.E.D.

2.4 Optimizing the Square Partition and the Row-in/Column-out Algorithms

The results of Sect. 2.3.2 can now be used to help us find the optimal way to transpose an arbitrary matrix with the square partition algorithm or the row-in/column-out algorithm.

More precisely, we shall minimize the memory requirements for a given number of data passes. This does not always minimize the number of IO operations, but the deviation from the minimum is small (see Sect. 2.3.2). Moreover, if the optimum with respect to memory is obtained without insertion of zero rows ($\bar{M} = M$), then the IO requirements are also minimized (Sect. 2.2.3). Finally, since the number of data passes needed for an $M \times N$ matrix is bounded above by $\lceil \log_2 M \rceil$ ($M \leq N$) or $\lceil \log_2 N \rceil + 1$ ($M > N$), the (approximate) optimum with respect to memory *and* IO can be found, given the relative costs of storage and IO operations.

Consider first the case $M \leq N$. Then, for given p, we can proceed as follows:

1) Set $m = \lceil M^{1/p} \rceil$. (Then $m = \min \{k; k^p \geq M\}$.)

2) Find the smallest product $\bar{M} = m_1 \cdot ... \cdot m_p \geq M$, with $m_1 \geq ... \geq m_p$, using $m_1 = m$ because $(m-1)^p < M$ according to 1.

3) Compute the number of high-speed locations needed, $\text{RM}_S(\bar{M})$, according to (2.25).

4) Search for a better product $\bar{M}' = m'_1 \cdot ... \cdot m'_p \geq M$, $m'_1 \geq ... \geq m'_p$, for which $m'_1 < \text{RM}_S(\bar{M})/N$. [We only have to consider these products since $\text{RM}_S(\bar{M}') \geq m'_1 N$]. Note that $m'_1 \geq m_1$ because $\bar{M}' \geq \bar{M}$.

5) If such an $\bar{M}'$ is found, replace $\bar{M}$ with $\bar{M}'$ and repeat 4.

The set of products in 4 is preferably searched in (increasing) lexicographic order. Moreover, from the choice of $\bar{M}$ and Theorem 3 it also follows that we only have to consider products $\bar{M}' > N$; that is, the search in 4 is over the set

$$\Omega = \{M' = m'_1 \cdot ... \cdot m'_p; m'_1 \geq ... \geq m'_p, \bar{M} > N, m'_1 \leq \text{RM}_S(\bar{M})/N\}. \tag{2.51}$$

The case $M > N$ is slightly more complicated. First, one considers the matrix as partitioned into $N \times N$ matrices and computes the optimal p product, $\bar{M}$, for these. If $\bar{M} \geq M$ one has the best p-step factorization. If, on the other hand $\bar{M} < M$, an additional step is needed to concatenate the rows into rows with M

Table 2.1. Memory requirements for transposition of a 620×1000 matrix. $RM_S(\bar{M})$ is always optimal

p	$\bar{M}$	$RM_S(\bar{M})$	Ω
2	25^2	25,000	$\emptyset$
3	$9^2 \cdot 8$	9,072	$\emptyset$
4	5^4	5,000	$\emptyset$
5	$4^4 \cdot 3$	4,096	$\{4^5\}$
6	3^6	3,645	$\emptyset$
7	$3^4 \cdot 2^3$	3,078	$\{3^{7-k}2^k; k=0,\ldots 2\}$
8	$3^3 \cdot 2^5$	3,024	$\{3^{8-k2k}; k=0,\ldots, 4\}$
9	$3 \cdot 2^8$	3,000	$\{3^{9-k}2^k; k=0,\ldots, 7\}$
10	2^{10}	2,048	$\emptyset$

elements. Discarding all zero elements one then needs $\max\{M, RM_S(\bar{M})\}$ storage locations for the $(p+1)$ step transposition. However, a better result can possibly be obtained if one directly applies a $(p+1)$ step transposition.

For instance, if $M=27$, $N=25$, and $p=2$, then $\bar{M}=5 \cdot 5$ and 125 storage locations are needed using 3 passes. But a 27×25 matrix can be transposed in 3 steps using 81 storage locations $(M=3 \cdot 3 \cdot 3)$. Observe that by Theorem 2, transpositions of the $N \times N$ matrices in $(p-1)$ steps followed by concatenation of rows never can give better results if $\bar{M} \geq M$ above.

In Table 2.1 we give an example of how this method works for a 620×1000 matrix.

2.5 An Example

To illustrate how the 4 algorithms presented in Sects. 2.2.3–6 work we apply them to a 6×6 matrix.

1) Floyd's method. $RM=12$, $IO=24$. Note that $A^{(1)}$, $B^{(0)}$, and $B^{(3)}$ are not constructed.

$$\begin{bmatrix} 00 & 01 & 02 & 03 & 04 & 05 \\ 10 & 11 & 12 & 13 & 14 & 15 \\ 20 & 21 & 22 & 23 & 24 & 25 \\ 30 & 31 & 32 & 33 & 34 & 35 \\ 40 & 41 & 42 & 43 & 44 & 45 \\ 50 & 51 & 52 & 53 & 53 & 55 \end{bmatrix}$$

a) $A = A^{(0)}$

$$\begin{bmatrix} 00 & 01 & 02 & 03 & 04 & 05 \\ 50 & 51 & 52 & 53 & 54 & 55 \\ 40 & 41 & 42 & 43 & 44 & 45 \\ 30 & 31 & 32 & 33 & 34 & 35 \\ 20 & 21 & 22 & 23 & 24 & 25 \\ 10 & 11 & 12 & 13 & 14 & 15 \end{bmatrix}$$

b) $A^{(1)}$, where $a_{i,j}^{(1)} = a_{-i,j}$

$$\begin{bmatrix} 00 & 01 & 02 & 03 & 04 & 05 \\ 51 & 52 & 53 & 54 & 55 & 50 \\ 42 & 43 & 44 & 45 & 40 & 41 \\ 33 & 34 & 35 & 30 & 31 & 32 \\ 24 & 25 & 20 & 21 & 22 & 23 \\ 15 & 10 & 11 & 12 & 13 & 14 \end{bmatrix}$$

c) $B^{(0)}$, where $b_{i,i+j}^{(0)} = a_{i,j}^{(1)}$

$$\begin{bmatrix} 00 & 10 & 02 & 12 & 04 & 14 \\ 51 & 01 & 53 & 03 & 55 & 05 \\ 42 & 52 & 44 & 54 & 40 & 50 \\ 33 & 43 & 35 & 45 & 31 & 41 \\ 24 & 34 & 20 & 30 & 22 & 32 \\ 15 & 25 & 11 & 21 & 13 & 23 \end{bmatrix} \qquad \begin{bmatrix} 00 & 10 & 20 & 30 & 04 & 14 \\ 51 & 01 & 11 & 21 & 55 & 05 \\ 42 & 52 & 02 & 12 & 40 & 50 \\ 33 & 43 & 53 & 03 & 31 & 41 \\ 24 & 34 & 44 & 54 & 22 & 32 \\ 15 & 25 & 35 & 45 & 13 & 23 \end{bmatrix}$$

d) $B^{(1)}$, columns with $j_0 = 1$ shifted 1 step

e) $B^{(2)}$, columns with $j_1 = 1$ shifted 2 steps

$$\begin{bmatrix} 00 & 10 & 20 & 30 & 40 & 50 \\ 51 & 01 & 11 & 21 & 31 & 41 \\ 42 & 52 & 02 & 12 & 22 & 32 \\ 33 & 43 & 53 & 03 & 13 & 23 \\ 24 & 34 & 44 & 54 & 04 & 14 \\ 15 & 25 & 35 & 45 & 55 & 05 \end{bmatrix} \qquad \begin{bmatrix} 00 & 10 & 20 & 30 & 40 & 50 \\ 01 & 11 & 21 & 31 & 41 & 51 \\ 02 & 12 & 22 & 32 & 42 & 52 \\ 03 & 13 & 23 & 33 & 43 & 53 \\ 04 & 14 & 24 & 34 & 44 & 54 \\ 05 & 15 & 25 & 35 & 45 & 55 \end{bmatrix}$$

f) $B^{(3)}$, columns with $j_2 = 1$ shifted 4 steps

g) A^1, where $a'_{i,j} = b^{(3)}_{i,i+j}$

2) The square method: $\bar{M} = 3 \cdot 2$, RM $= 18$, IO $= 24$. We have indicated how the rows are grouped together and shown how one group of rows maps in step 2.

$$\begin{bmatrix} 00 & 01 & 02 & 03 & 04 & 05 \\ 10 & 11 & 12 & 13 & 14 & 15 \\ 20 & 21 & 22 & 23 & 24 & 25 \\ 30 & 31 & 32 & 33 & 34 & 35 \\ 40 & 41 & 42 & 43 & 44 & 45 \\ 50 & 51 & 52 & 53 & 54 & 55 \end{bmatrix} \rightarrow \begin{bmatrix} 00 & 10 & 20 & 03 & 13 & 23 \\ 01 & 11 & 21 & 04 & 14 & 24 \\ 02 & 12 & 22 & 05 & 15 & 25 \\ 30 & 40 & 50 & 33 & 43 & 53 \\ 31 & 41 & 51 & 34 & 44 & 54 \\ 32 & 42 & 52 & 35 & 45 & 55 \end{bmatrix} \rightarrow \begin{bmatrix} 00 & 10 & 20 & 30 & 40 & 50 \\ - & & & & & \\ - & & & & & \\ 03 & 13 & 23 & 33 & 43 & 53 \\ - & & & & & \\ - & & & & & \end{bmatrix}$$

3) The rectangular method: $\bar{M} = 2 \cdot 3$, $\bar{N} = 3 \cdot 2$, RM $= 18$ (RM $= 12$ also possible, see [2.6]), IO $= 30$.

$$\begin{bmatrix} 00 & 01 & 02 & 03 & 04 & 05 \\ 10 & 11 & 12 & 13 & 14 & 15 \\ 20 & 21 & 22 & 23 & 24 & 25 \\ 30 & 31 & 32 & 33 & 34 & 35 \\ 40 & 41 & 42 & 43 & 44 & 45 \\ 50 & 51 & 52 & 53 & 54 & 55 \end{bmatrix} \rightarrow \begin{bmatrix} 00 & 10 & 03 & 13 \\ 01 & 11 & 04 & 14 \\ 02 & 12 & 05 & 15 \\ 20 & 30 & 23 & 33 \\ 21 & 31 & 24 & 34 \\ 22 & 32 & 25 & 35 \\ 40 & 50 & 43 & 53 \\ 41 & 51 & 44 & 54 \\ 42 & 52 & 45 & 55 \end{bmatrix} \rightarrow \begin{bmatrix} 00 & 10 & 20 & 30 & 40 & 50 \\ & & & & & \\ 03 & 13 & 23 & 33 & 43 & 53 \end{bmatrix}$$

4) The rectangular method: $\bar{M} = 3 \cdot 2$, $\bar{N} = 2 \cdot 3$, RM = 18, IO = 20.

$$
\begin{bmatrix}
00 & 01 & 02 & 03 & 04 & 05 \\
10 & 11 & 12 & 13 & 14 & 15 \\
20 & 21 & 22 & 23 & 24 & 25 \\
30 & 31 & 32 & 33 & 34 & 35 \\
40 & 41 & 42 & 43 & 44 & 45 \\
50 & 51 & 52 & 53 & 54 & 55
\end{bmatrix}
\rightarrow
\begin{bmatrix}
00 & 10 & 20 & 02 & 12 & 22 & 04 & 14 & 24 \\
01 & 11 & 21 & 03 & 13 & 23 & 05 & 15 & 25 \\
30 & 40 & 50 & 32 & 42 & 52 & 34 & 44 & 54 \\
31 & 41 & 51 & 33 & 43 & 53 & 35 & 45 & 55
\end{bmatrix}
$$

$$
\rightarrow
\begin{bmatrix}
00 & 10 & 20 & 30 & 40 & 50 \\
02 & 12 & 22 & 32 & 42 & 52 \\
04 & 14 & 24 & 34 & 44 & 54
\end{bmatrix}
$$

5) The row-in/column-out method: $\bar{M} = 3 \cdot 2$, RM = 18, IO = 24.

$$
\begin{bmatrix}
00 & 01 & 02 & 03 & 04 & 05 \\
10 & 11 & 12 & 13 & 14 & 15 \\
20 & 21 & 22 & 23 & 24 & 25 \\
30 & 31 & 32 & 33 & 34 & 35 \\
40 & 41 & 42 & 43 & 44 & 45 \\
50 & 51 & 52 & 53 & 54 & 55
\end{bmatrix}
\rightarrow
\begin{bmatrix}
00 & 10 & 20 & 01 & 11 & 21 \\
02 & 12 & 22 & 03 & 13 & 23 \\
04 & 14 & 24 & 05 & 15 & 25 \\
30 & 40 & 50 & 31 & 41 & 51 \\
32 & 42 & 52 & 33 & 43 & 53 \\
34 & 44 & 54 & 35 & 45 & 55
\end{bmatrix}
$$

$$
\rightarrow
\begin{bmatrix}
00 & 10 & 20 & 30 & 40 & 50 \\
01 & 11 & 21 & 31 & 41 & 51
\end{bmatrix}
$$

2.6 Anderson's Direct Method for Computing the FFT in Higher Dimensions

An alternative solution to the entire problem of computing the DFT of a matrix stored rowwise on a random access device is suggested by *Anderson* [2.1]. Because the outer sum, (2.2b), is a DFT for a fixed l, one can obviously use the FFT to compute it. The basic idea behind the FFT is that the DFT of an array can be computed as an iterated sum, where each sum in the iteration is itself a DFT. If, in particular, the array size, M, is a power of 2, each partial sum can be a 2 point DFT. Then (see, e.g., [2.16] or [2.17]) at step i, $i = 1, ..., \log_2 M$, one transforms the two points with indices m and m', for which the binary expansions differ exactly at position i. The final array will then appear in bit-reversed order, but the initial array can be rearranged so that the resulting values are in natural order. (See also [2.18].)

Observing that the sum in (2.2b) is computed for all l, we can proceed as follows. At step i, the row pairs for which the indices differ precisely at binary position i are successively read into high-speed storage, where the two point DFT is computed in place for all the columns. These modified rows are then written back to their old positions. After $\log_2 M$ steps we obtain the desired transform, but with the bit-reversed order. This is no problem because one can use a proper order key at input or during future processing. Irrespective of how the bit reversal is handled, one can observe that the addressing of the rows is completely analogous to the addressing in the square partition method (and in Floyd's algorithm) for transposition.

The analogy between the two methods extends to the possible generalizations that exist. In both cases, one can reduce the the IO time by treating more than two rows simultaneously. In fact, if the matrix contains $M \times N$ elements and one can store 2^m rows in high-speed memory, then the number of IO operations will be

$$IO_A = 2 \cdot M \cdot \left\lceil \frac{\log_2 M}{m} \right\rceil . \tag{2.52}$$

If M and N are powers of 2, then IO_A coincides with IO_S and IO_F. Furthermore, the method generalizes to arbitrary factorizations of M in the same manner as the FFT and the analogy with the transposition methods described in Sect. 2.2 still holds. Consequently, the performances measured in terms of IO operations are the same in this case as well.

The direct approach is conceptually simple, in that it is based on properties of the fast algorithm (e.g., the FFT) itself. It also has the advantage that it works in place on arbitrary matrices. However, there are also disadvantages with the method, that will be discussed in the next section.

2.7 Discussion

Two approaches for computing separable 2-D transforms for matrices accessible row- or columnwise have been presented. One of the methods is based on matrix transposition, the other on properties of "fast" transforms, such as the FFT. The methods have been shown to be essentially equivalent in performance when the matrix dimensions are highly composite numbers. (We assume that no actual transposition is done in high-speed storage, so that the processing times are approximately the same.) In such cases the direct approach can be used for in place transformation of arbitrary matrices. It is, however, less modular in its design and therefore the more general matrix transposition approach often gives better performance in other cases, as will be shown shortly. The modularity of the transposition method also implies that it can be combined with the use of fast transform hardware.

The advantage of the modularity is especially evident in the following case. Suppose that we want to compute the FFT of an $M \times N$ matrix and that M contains a large prime factor. The summations in the FFT will then be the same for the two methods, but the required amount of high-speed storage may be drastically reduced if transposition is done using a suitable embedding. Padding the arrays with zeros in the FFT computations in the direct method may not be desirable, because it changes the transform values. Consider for instance, the example in Table 2.1. The direct approach will require 2 data passes and 31,000 high-speed storage locations, which can be compared to the 25,000 locations needed to transpose in 2 passes. (Observe that the 1-D FFT computations can be intermingled with the transposition so that 2 passes are sufficient.) Futhermore, Table 2.1 shows that the transposition method can be applied with much less memory, at the expense of increased IO time. Conversely, there are also cases where memory can be traded for IO time.

Let us mention some other, less important differences between the methods. Assume that we want to compute the transform and the inverse transform of an $M \times N$ matrix and that M and N are highly composite, so that nothing is gained by padding the matrix with zeros. Assume that KN elements fit into high-speed storage and that KN/M is integer, then,

$$IO_S = 2 \left(\left\lceil \frac{\log_2 M}{\log_2 K} \right\rceil M + \left\lceil \frac{\log_2 N}{\log_2 KN/M} \right\rceil N \right) \tag{2.53}$$

and

$$IO_A = 4 \left\lceil \frac{\log_2 M}{\log_2 K} \right\rceil M . \tag{2.54}$$

It can be shown from these expressions that $IO_S < IO_A$ if $M < N$ and that $IO_S > IO_A$ if $M > N$. If, for example, $M = 2^{10}$, $N = 2^8$ and $K = 2^3$ then $IO_S = 48 \cdot 2^8$ whereas $IO_A = 64 \cdot 2^8$, but if $M = 2^8$ and $N = 2^{10}$ then $IO_S = 22.2^8$ and $IO_A = 12 \cdot 2^8$. It can also be noted that the minimum amount of storage required to compute a transform pair, regardless of the IO time, is smaller for the direct approach if $M > N$.

The four algorithms for matrix transposition that have been presented are all optimal in the case of a matrix whose dimensions are powers of 2. In this case they can also all be implemented using shift and masking operations only. However, they differ slightly when applied to arbitrary rectangular matrices.

Floyd's algorithm can handle any square matrix using only shifts and masking and extends to rectangular matrices in the manner described in Sect. 2.2.3. However, the other three algorithms do not have this "binary" property, because they utilize expansions of the form (2.15). On the other hand they can perform the transposition in fewer IO operations, using slightly more memory.

These three algorithms also require an optimization step, which for the square partition and the row-in/column-out algorithms is very simple, but somewhat more complicated for the rectangular partition algorithm.

The latter algorithm is also conceptually more complex, because the number of rows as well as the row lengths vary during the transposition, even when the matrix dimensions are composite numbers.

In general, the square partition and the row-in/column-out algorithms, which are simpler, give (almost) as good performance. They are therefore preferable, unless the implementation has to be done with shifts and maskings, when *Floyd*'s algorithm can be used.

References

2.1   G.L.Anderson: IEEE ASSP-**28**, 280–284 (1980)
2.2   N.M.Brenner: IEEE AU-**17**, 128–132 (1969)
2.3   B.R.Hunt: Proc. IEEE **60**, 884–887 (1972)
2.4   R.C.Singleton: IEEE AU-**15**, 91–98 (1967)
2.5   H.L.Buijs: Appl. Opt. **8**, 211–212 (1969)
2.6   J.O.Eklundh: "Efficient Matrix Transposition with Limited High-Speed Storage"; FOA Reports, Vol. 12, No. 1, National Defence Research Institute, Stockholm, Sweden (1978)
2.7   J.O.Eklundh: IEEE C-**21**, 801–803 (1972)
2.8   H.K.Ramapriyan: IEEE C-**24**, 1221–1226 (1976)
2.9   R.E.Twogood, M.P.Ekstrom: IEEE C-**24**, 950–952 (1976)
2.10 R.W.Floyd: "Permuting Information in Idealized Two-Level Storage", in *Complexity of Computer Computations*, ed. by R.E.Miller, J.W.Thatcher (Plenum Press, New York 1972) pp. 105–109
2.11 H.S.Stone: IEEE C-**20**, 153–161 (1971)
2.12 D.E.Knuth: *The Art of Computer Programming*, Vol. 3 (Addison-Wesley, Reading, MA 1973) pp. 7, 573
2.13 L.G.Delcaro, G.L.Sicuranza: IEEE C-**23**, 967–970 (1974)
2.14 U.Schumann: Angew. Informatik **14**, 213–216 (1972)
2.15 U.Schumann: IEEE C-**22**, 542–543 (1973)
2.16 J.W.Cooley, J.W.Tukey: Math. Comput. **19**, 297–301 (1965)
2.17 W.T.Cochran et al.: IEEE AU-**15**, 45–55 (1967)
2.18 G.E.Rivard: IEEE ASSP-**25**, 250–252 (1977)

3. Two-Dimensional Convolution and DFT Computation

H. J. Nussbaumer

With 8 Figures

The main objective of this chapter is to develop fast algorithms for the computation of two-dimensional convolutions and discrete Fourier transforms (DFTs). These algorithms are based upon an efficient two-dimensional to one-dimensional mapping, using the basic properties of rings of polynomials, and they significantly reduce the number of arithmetic operations required to compute convolutions and DFTs, when compared to more conventional methods based on the fast Fourier transform (FFT).

   In this chapter, we first give a brief overview of those parts of polynomial algebra which relate to the computation of convolutions and DFTs. We then introduce new transforms, called polynomial transforms, and show that these transforms, which can be viewed as DFTs defined in fields of polynomials, provide an efficient tool for computing two-dimensional filtering processes. We also discuss the use of polynomial transforms for the computation of two-dimensional DFTs and we show that these transforms can be used in combination with the Winograd algorithm or the FFT to further reduce the number of arithmetic operations.

3.1 Convolutions and Polynomial Algebra

Many of the newest techniques for the fast computation of convolutions and DFTs such as the Winograd Fourier transform algorithm [3.1], the Agarwal–Cooley fast convolution algorithm [3.2] and polynomial transforms [3.3–5] are based on polynomial algebra. Since polynomial algebra has not yet been widely used in electrical engineering, except for coding, we give here a brief intuitive overview of its most important aspects concerning convolutions. For more details, the reader can reference to any textbook on modern algebra [3.6].

   We consider first a discrete convolution process defined by

$$y_m = \sum_{l=0}^{L-1} h_l x_{m-l} \,. \tag{3.1}$$

The direct computation of this finite discrete convolution requires L multiplications per output sample. In most fast computation algorithms, the infinite input sequence $\{x_n\}$ is divided into consecutive blocks of N samples and y_m is calculated as a sum of aperiodic convolutions of each of these blocks with the sequence $\{h_l\}$ having L points.

Computing these aperiodic convolutions is equivalent to determining the coefficients of the product $Y(Z)$ of two polynomials $H(Z)$ and $X(Z)$ with

$$H(Z) = \sum_{l=0}^{L-1} h_l Z^l \tag{3.2}$$

$$X(Z) = \sum_{n=0}^{N-1} x_n Z^n \tag{3.3}$$

$$Y(Z) = H(Z)X(Z) = \sum_{m=0}^{L+N-2} y_m Z^m \tag{3.4}$$

where h_l and x_n are elements of some field F, usually the field of real or complex numbers. The polynomial representation of a convolution is obviously more complex than the regular representation. We shall see, however, that using polynomials provides an efficient tool for reducing the complexity of computing convolutions and polynomial products. In order to do this, we need first to introduce residue polynomials and the Chinese remainder theorem.

3.1.1 Residue Polynomials

Assuming polynomials with coefficients defined in a field, a polynomial $P(Z)$ divides a polynomial $H(Z)$ if a polynomial $D(Z)$ can be found such that $H(Z) = P(Z)D(Z)$. $H(Z)$ is said to be irreducible if its only divisors $P(Z)$ are of degree equal to zero. If $P(Z)$ is not a divisor of $H(Z)$, the division of $H(Z)$ by $P(Z)$ will produce a residue $R(Z)$

$$H(Z) = P(Z)D(Z) + R(Z). \tag{3.5}$$

It can be shown easily that this representation is unique. All polynomials that yield the same residue $R(Z)$ when divided by $P(Z)$ are said to be congruent modulo $P(Z)$. This congruence relation is denoted

$$R(Z) \equiv H(Z) \quad \text{modulo } P(Z). \tag{3.6}$$

The computation of $R(Z)$ is greatly simplified if $P(Z)$ is the product of d polynomials having no common factors (usually called relatively prime polynomials by analogy with relatively prime numbers)

$$P(Z) = \prod_{i=1}^{d} P_i(Z). \tag{3.7}$$

In this case, $R(Z)$ can be expressed uniquely as a function of the polynomials $H_i(Z)$ obtained by reducing $H(Z)$ modulo the various polynomials $P_i(Z)$. This expression, which is an extension to rings of polynomials of the Chinese remainder theorem [3.7], known in number theory, is given by

$$R(Z) \equiv H(Z) \quad \text{modulo } P(Z) \equiv \sum_{i=1}^{d} S_i(Z) H_i(Z) \quad \text{modulo } P(Z) \tag{3.8}$$

where, for each value u of i,

$$S_u(Z) \equiv 0 \quad \text{modulo } P_i(Z), \quad i \neq u \tag{3.9}$$

$$\equiv 1 \quad \text{modulo } P_u(Z)$$

and

$$S_u(Z) \equiv \left\{ \left[\prod_{\substack{i=1 \\ i \neq u}}^{d} P_i(Z) \right] \middle/ \left\{ \left[\prod_{\substack{i=1 \\ i \neq u}}^{d} P_i(Z) \right] \text{ modulo } P_u(Z) \right\} \right\} \text{ modulo } P(Z). \tag{3.10}$$

The relations (3.8, 9) can be verified easily by reducing (3.8) modulo the various polynomials $P_u(Z)$. Since $[H(Z)$ modulo $P(Z)]$ modulo $P_i(Z) \equiv H_i(Z)$, the relation (3.8) holds for any $P_u(Z)$. Similarly, reducing $S_u(Z)$, defined by (3.10), modulo $P_u(Z)$ yields (3.9) provided $\prod_{\substack{i=1 \\ i \neq u}}^{d} P_i(Z) \not\equiv 0$ modulo $P_u(Z)$, this last condition being insured by the fact that the polynomials $P_u(Z)$ are relatively prime. $S_u(Z)$ can be constructed easily by using a polynomial version of Euclid's algorithm.

3.1.2 Convolution and Polynomial Product Algorithms in Polynomial Algebra

The Chinese remainder theorem plays a central role in the computation of convolutions because it allows us to replace the computation of the product $H(Z)X(Z)$ of two large polynomials modulo $P(Z)$ by that of d products $H_i(Z)X_i(Z)$ modulo $P_i(Z)$ of much smaller polynomials. Here, $X_i(Z) = X(Z)$ modulo $P_i(Z)$. We now show how the Chinese remainder theorem can be used to specify the minimum number of multiplications required for computing a convolution or a polynomial product, and how these lower bounds can be achieved in practice.

Theorem 3.1. (*The Cook–Toom algorithm* [3.2].) The aperiodic convolution (3.4) can be computed in $L + N - 1$ general multiplications.

A constructive proof is given by noting that $Y(Z)$ is of degree $L + N - 2$. Therefore, $Y(Z)$ is unchanged if it is defined modulo any polynomial $P(Z)$ of degree equal to $L+N-1$

$$Y(Z) \equiv H(Z)X(Z) \quad \text{modulo } P(Z). \tag{3.11}$$

Let us assume that $P(Z)$ is chosen to be

$$P(Z) = \prod_{i=1}^{L+N-1} (Z-a_i) \tag{3.12}$$

where the a_i are $L+N-1$ distinct numbers in the field F of coefficients. $Y(Z)$ can be computed by reducing $H(Z)$ and $X(Z)$ modulo $(Z-a_i)$, which is equivalent to substituting a_i for Z in $H(Z)$ and $X(Z)$. Then $Y(Z)$ is calculated in $L+N-1$ general multiplications by performing the $L+N-1$ products $H(a_i)X(a_i)$ and reconstructing $Y(Z)$ by the Chinese remainder theorem. This completes the proof of Theorem 3.1.

Note that the reductions modulo $(Z-a_i)$ and the Chinese remainder reconstruction imply multiplications by scalars which are powers of a_i. For short convolutions, the set of $L+N-1$ distinct a_i can be chosen to be simple integers such as 0, $+1$, -1 so that these multiplications are either trivial or reduce to a few additions. For larger convolutions, however, the requirement that all a_i be distinct leads to the choice of larger numbers so that these scalar multiplications must either be treated as general multiplications or require a large number of additions. Thus, the Cook–Toom algorithm is in practice used only for short convolutions. For larger convolutions, a better balance between minimizing the number of multiplications and minimizing the number of additions can be achieved by using transform approaches which we show now to be closely related to the Cook–Toom algorithm.

In the Cook–Toom algorithm, the result polynomial $Y(Z)$ of degree $L+N-2$ is reconstructed from $L+N-1$ numerical values $Y(a_i)$ obtained by substituting a_i for Z in $H(Z)$ and $X(Z)$. Thus, the Cook–Toom algorithm can be viewed as a Lagrange interpolation process [3.2]. The choice of the $L+N-1$ distinct a_i and of the field of coefficients F is arbitrary. If the various a_i are chosen to be successive powers of a number W and if the field F is such that 1, W, $W^2, \ldots, W^{L+N-2}$ are all distinct, the Cook–Toom algorithm reduces to computing an aperiodic convolution with transforms having the DFT structure: these transforms would be, for instance, Mersenne or Fermat transforms [3.8, 9] if $W=2$ and if F is the ring of numbers modulo a Mersenne number $(2^q-1, q$ prime$)$ or a Fermat number $(2^q+1, q=2^t)$.

When F is the field of complex numbers and $W=\exp[-j2\pi/(L+N-1)]$, $j=\sqrt{-1}$, using the Cook–Toom algorithm is equivalent to computing the aperiodic convolution with DFTs. In this case, $P(Z)$ reduces to

$$P(Z) = \prod_{i=0}^{L+N-2} (Z-W^i) = Z^{L+N-1}-1 \tag{3.13}$$

and the aperiodic convolution of the two sequences $\{h_l\}$ of L points and $\{x_n\}$ of N points is computed as a circular convolution of two extended sequences of $L+N-1$ points obtained by appending $N-1$ zeros to the sequence $\{h_l\}$ and $L-1$ zeros to the sequence $\{x_n\}$. Thus, the familiar method of computing aperiodic convolutions by DFTs and the overlap-add technique [3.10] is closely related to the Cook–Toom algorithm.

In Theorem 3.1, we have seen that an aperiodic convolution of $L+N-2$ points could be computed with $L+N-1$ multiplications by performing a polynomial multiplication modulo a polynomial $P(Z)$, of degree $D=L+N-1$, chosen to be the product of D first degree polynomials in the field of coefficients F. The following important theorem due to *Winograd* [3.11] generalizes these results to the case of polynomial products modulo any polynomial $P(Z)$.

Theorem 3.2. A polynomial product $Y(Z) \equiv H(Z)X(Z)$ modulo $P(Z)$ is computed with a minimum of $2D-d$ general multiplications, where D is the degree of $P(Z)$ and d is the number of irreducible factors $P_i(Z)$ of $P(Z)$ over the field F (including 1).

A constructive proof of this theorem is given by using the Chinese remainder theorem. Since $P(Z)$ is the product of d irreducible polynomials $P_i(Z)$, $Y(Z)$ can be computed by reducing $H(Z)$ and $X(Z)$ modulo $P_i(Z)$, evaluating the d polynomial products $Y_i(Z) \equiv H_i(Z)X_i(Z)$ modulo $P_i(Z)$ corresponding to the reduced polynomials $H_i(Z)$, $X_i(Z)$ and reconstructing $Y(Z)$ from $Y_i(Z)$ by the Chinese remainder theorem. As for Theorem 3.1, the multiplications by scalars involved in the reductions and Chinese remainder reconstruction are not counted and the only general multiplications are those corresponding to the products $H_i(Z)X_i(Z)$. Let D_i be the degree of $P_i(Z)$ with $D = \sum_{i=1}^{d} D_i$. Since $P_i(Z)$ is of degree D_i, $H_i(Z)$ and $X_i(Z)$ are of degree D_i-1, and their product can be calculated with $2D_i-1$ multiplications by using Theorem 3.1

$$Y_i(Z) \equiv [H_i(Z)X_i(Z) \quad \text{modulo } Q_i(Z)] \quad \text{modulo } P_i(Z) \tag{3.14}$$

where $Q_i(Z)$ is a polynomial of degree $2D_i-1$ chosen to be the product of $2D_i-1$ first degree polynomials in the field F. Thus, the number of general multiplications is given by $\sum_{i=1}^{d}(2D_i-1) = 2D-d$. This completes the proof of Theorem 3.2.

An important consequence of Theorem 3.2 concerns the computation of circular convolutions. A circular convolution of two sequences having D points can be viewed in polynomial notation as a polynomial product modulo (Z^D-1). We have already seen that if F is the field of complex numbers, (Z^D-1) factors into D first degree polynomials $(Z-W^i)$ and that the circular convolution is computed by DFTs with D general multiplications. Unfortunately, the W^i's are irrational and complex so that the multiplications by scalars involved in computing the DFTs must be considered as general multiplications.

When F is the field of rational numbers, $(Z^D - 1)$ factors into a product of polynomials whose coefficients are rational numbers. These polynomials are called *cyclotomic polynomials* [3.7]. Cyclotomic polynomials are irreducible over the field of rational numbers. Their roots are complex and a subset of the set W^0, W^1, ..., W^{D-1}, with $W = \exp(-j2\pi/D)$, $j = \sqrt{-1}$.

It can be shown that, for a given D, the number d of distinct cyclotomic polynomials factors of $(Z^D - 1)$ is equal to the number of divisors of D, including 1 and D. Furthermore, the degree of each cyclotomic polynomial corresponding to a given D_i divisor of D, is equal to Euler's totient function $\varphi(D_i)$ [3.7]. Thus, for circular convolutions, Theorem 3.2 reduces to Theorem 3.3.

Theorem 3.3. A circular convolution of D points is computed with a minimum of $2D - d$ general multiplications, where d is the number of divisors of D, including 1 and D.

The computation of circular convolutions by Theorem 3.2 is greatly simplified by the fact that the coefficients of cyclotomic polynomials are simple integers. Moreover, these integers can be only 0, $+1$ or -1, except for very large cyclotomic polynomials [3.2]. In this case, the reductions and Chinese remainder reconstruction can be implemented with a small number of additions as illustrated by the simple example of a circular convolution of 5 points. In this example, the two only divisors of 5 are 1 and 5. Thus, $(Z^5 - 1)$ factors into two cyclotomic polynomials with $(Z^5 - 1) = (Z - 1)$ $(Z^4 + Z^3 + Z^2 + Z + 1)$. Reducing the input polynomials $H(Z)$ and $X(Z)$ modulo $(Z - 1)$ amounts to a simple substitution of 1 for Z in $H(Z)$ and $X(Z)$ and is therefore done with 4 additions. The reductions modulo $Z^4 + Z^3 + Z^2 + Z + 1$ are also done with 4 additions by substracting the term of degree 4 to all other terms of the input polynomials, since $Z^4 \equiv -1 - Z - Z^2 - Z^3$. The multiplication modulo $(Z - 1)$ is a simple scalar multiplication and the multiplication modulo $(Z^4 + Z^3 + Z^2 + Z + 1)$ is done with 7 multiplications by using Theorem 3.1. The Chinese remainder reconstruction can be viewed as the inverse of the reductions and is done with 8 additions. Thus, a circular convolution of 5 points is computed with only 8 multiplications and a limited number of additions for reductions and Chinese remainder reconstruction.

3.2 Two-Dimensional Convolution Using Polynomial Transforms

Direct computation of a two-dimensional nonrecursive filtering process applied to an array of dimension $N \times N$ corresponds to a number of operations proportional to $N^2 M^2$ (where $M \times M$ is the size of the convolution window) and therefore requires a large computing power, even for relatively small convolutions. This has stimulated a number of efforts to find more efficient methods for implementing two-dimensional nonrecursive digital filters.

In most of these methods, the digital nonrecursive filtering process is evaluated by computing a series of circular convolutions obtained by segmenting the input sequences and appending a suitable number of zeros to these segmented sequences. The final result of the digital filtering process is then obtained by the overlap-add or overlap-save method [3.10]. With this approach, the bulk of the computation corresponds to the evaluation of the circular convolutions, which is usually greatly simplified by using FFTs [3.12], number theoretic transforms (NTT) [3.8, 9] or the Agarwal–Cooley method [3.2]. When compared with direct computation, these various techniques allow a substantial reduction in number of operations but are generally suboptimal for evaluating two-dimensional convolutions. Computation by FFTs operates on complex numbers, introduces a significant amount of roundoff errors, requires some means of processing trigonometric functions and still corresponds to a large number of multiplications. NTTs can be computed without multiplications and eliminate quantization errors in the computation of convolutions. However, these transforms are severely restricted in word length and transform length and use modular arithmetic which is usually not implemented very efficiently in general purpose computers. The Agarwal–Cooley method, based on a nesting of several short convolutions, is attractive for one-dimensional convolutions because it does not use modular arithmetic nor require any manipulation of trigonometric functions. Unfortunately, this method is not very efficient for two-dimensional convolutions.

We shall show here that the computation of two-dimensional convolutions can be greatly simplified by using polynomial transforms [3.3–5] to map two-dimensional convolutions into one-dimensional convolutions and polynomial products. Polynomial transforms are defined in fields of polynomials and have the circular convolution property. These transforms are computed in ordinary arithmetic, without multiplications and, when used with efficient short convolution and polynomial product algorithms, allow one to minimize the processing load required for evaluating two-dimensional convolutions.

3.2.1 Polynomial Transforms

Let $y_{u,l}$ be a two-dimensional circular convolution of dimension $N \times N$, with

$$y_{u,l} = \sum_{m=0}^{N-1} \sum_{n=0}^{N-1} h_{n,m} x_{u-n,\, l-m} \qquad u,l = 0, ..., N-1. \tag{3.15}$$

In order to uncover the underlying simplifications that can be made in computing this convolution, we need a representation in polynomial algebra.

This is done by noting that (3.15) can be viewed as a one-dimensional polynomial convolution with

$$Y_l(Z) \equiv \sum_{m=0}^{N-1} H_m(Z)X_{l-m}(Z) \quad \text{modulo } (Z^N - 1) \tag{3.16}$$

$$H_m(Z) = \sum_{n=0}^{N-1} h_{n,m}Z^n, \quad m=0,\ldots,N-1 \tag{3.17}$$

$$X_r(Z) = \sum_{s=0}^{N-1} x_{s,r}Z^s, \quad r=0,\ldots,N-1, \tag{3.18}$$

where $y_{u,l}$ is obtained from the N polynomials $Y_l(Z)$ by taking the coefficients of Z^u in $Y_l(Z)$ with

$$Y_l(Z) = \sum_{u=0}^{N-1} y_{u,l}Z^u, \quad l=0,\ldots,N-1. \tag{3.19}$$

We assume first that N is an odd prime, with $N=q$. In this case, as shown in Sect. 3.1, $(Z^q - 1)$ is the product of two cyclotomic polynomials

$$Z^q - 1 = (Z-1)P(Z) \tag{3.20}$$

$$P(Z) = Z^{q-1} + Z^{q-2} + \ldots + 1. \tag{3.21}$$

Since $Y_l(Z)$ is defined modulo $(Z^q - 1)$, it can be computed by reducing $H_m(Z)$ and $X_r(Z)$ modulo $(Z-1)$ and $P(Z)$, the factors of $(Z^q - 1)$, computing the polynomial convolutions $Y_{1,l}(Z) \equiv Y_l(Z)$ modulo $P(Z)$ and $Y_{2,l} \equiv Y_l(Z)$ modulo $(Z-1)$ on the reduced polynomials and reconstructing $Y_l(Z)$ by the Chinese remainder theorem

$$Y_l(Z) \equiv S_1(Z) Y_{1,l}(Z) + S_2(Z) Y_{2,l} \quad \text{modulo } (Z^q - 1) \tag{3.22}$$

$$\begin{cases} S_1(Z) \equiv 1 & S_2(Z) \equiv 0 \quad \text{modulo } P(Z) \\ S_1(Z) \equiv 0 & S_2(Z) \equiv 1 \quad \text{modulo } (Z-1) \end{cases} \tag{3.23}$$

with

$$S_1(Z) = [q - P(Z)]/q \tag{3.24}$$

and

$$S_2(Z) = P(Z)/q. \tag{3.25}$$

Computing $Y_l(Z)$ is therefore replaced by the simpler problem of computing $Y_{1,l}(Z)$ and $Y_{2,l}$. The calculation of $Y_{2,l}$ is particularly easy because $Y_{2,l}$ is defined modulo $(Z-1)$. Thus, $Y_{2,l}$ is the convolution product of the scalars $H_{2,m}$ and $X_{2,r}$ obtained by substituting 1 for Z in $H_m(Z)$ and $X_r(Z)$

$$Y_{2,l} = \sum_{m=0}^{q-1} H_{2,m} X_{2,l-m}, \quad l=0,\ldots,q-1 \tag{3.26}$$

$$H_{2,m} = \sum_{n=0}^{q-1} h_{n,m} \quad X_{2,r} = \sum_{s=0}^{q-1} x_{s,r}. \tag{3.27}$$

The most difficult part of the computation of $Y_l(Z)$ is the calculation of $Y_{1,l}(Z)$. In order to simplify this calculation, we introduce a transform $\bar{H}_k(Z)$, defined modulo $P(Z)$ with

$$\bar{H}_k(Z) \equiv \sum_{m=0}^{q-1} H_{1,m}(Z) Z^{mk} \text{ modulo } P(Z) \tag{3.28}$$

$$H_{1,m}(Z) \equiv H_m(Z) \quad \text{modulo } P(Z)$$

$$k=0,\ldots,q-1.$$

We call this transform, which has the same structure as a DFT, but with complex exponentials replaced by Z, a *polynomial transform*. We define similarly an inverse transform by

$$H_{1,l}(Z) \equiv \frac{1}{q} \sum_{k=0}^{q-1} \bar{H}_k(Z) Z^{-lk} \text{ modulo } P(Z) \tag{3.29}$$

$$l=0,\ldots,q-1.$$

Polynomial transforms are computed with multiplications by powers of Z and additions. Polynomial additions are performed by adding separately the words corresponding to each coefficient of Z. Multiplications by powers of Z are simplified by noting that $Z^q \equiv 1$ modulo $P(Z)$. Thus, $H_{1,m}(Z) Z^{mk}$ modulo $P(Z) \equiv [H_{1,m}(Z) Z^{mk}$ modulo $(Z^q - 1)]$ modulo $P(Z)$. The multiplication of $H_{1,m}(Z)$ by Z^{mk} modulo $(Z^q - 1)$ is therefore a simple rotation by $[(mk)$ modulo $q]$ points of the words $(h_{0,m} - h_{q-1,m})$, $(h_{1,m} - h_{q-1,m}), \ldots, (h_{q-2,m} - h_{q-1,m})$, 0 of $H_{1,m}(Z)$ within a q words polynomial. Since $Z^{q-1} \equiv -Z^{q-2} - Z^{q-3} \ldots -1$ modulo $P(Z)$, the reduction modulo $P(Z)$ is done by substracting the coefficient of Z^{q-1} to all other coefficients of Z. Thus, polynomial transforms are computed with simple additions, without any restrictions on the arithmetic since the field F of the coefficients can be chosen at will.

We show now that polynomial transforms share with DFTs and NTTs the circular convolution property and can therefore be used to simplify the computation of convolutions. This can be seen by calculating the transforms $\bar{H}_k(Z)$ and $\bar{X}_k(Z)$ of $H_{1,m}(Z)$ and $X_{1,r}(Z)$ by (3.28), multiplying $\bar{H}_k(Z)$ by $\bar{X}_k(Z)$ modulo $P(Z)$ and computing the inverse transform $Q_l(Z)$ of $\bar{H}_k(Z)\bar{X}_k(Z)$

$$Q_l(Z) \equiv \sum_{m=0}^{q-1} \sum_{r=0}^{q-1} H_{1,m}(Z)X_{1,r}(Z)\frac{1}{q}\sum_{k=0}^{q-1} Z^{pk} \text{ modulo } P(Z) \tag{3.30}$$

with $p=m+r-l$. Let $S = \sum_{k=0}^{q-1} Z^{pk}$. For $p \not\equiv 0$ modulo q, the set of exponents pk modulo q is a simple permutation of the integers $0, 1, ..., q-1$. Thus, $S \equiv \sum_{k=0}^{q-1} Z^k \equiv P(Z) \equiv 0$ modulo $P(Z)$. For $p \equiv 0$ modulo q, $S=q$. Thus, the only non zero case corresponds to $p \equiv 0$ or $r \equiv l-m$ modulo q and $Q_l(Z)$ reduces to a circular polynomial convolution

$$Q_l(Z) \equiv Y_{1,l}(Z) \equiv \sum_{m=0}^{q-1} H_{1,m}(Z)X_{1,l-m}(Z) \quad \text{modulo } P(Z). \tag{3.31}$$

Under these conditions, $Y_{1,l}(Z)$ is computed with three polynomial transforms and q polynomial multiplications $\bar{H}_k(Z)\bar{X}_k(Z)$ defined modulo $P(Z)$. In many digital filtering applications, one of the input sequences is constant. Its transform can then be precomputed and only two polynomial transforms are required. In this case, the Chinese remainder reconstruction can also be simplified by noting with (3.10) that

$$S_1(Z) \equiv (Z-1)/[(Z-1) \text{ modulo } P(Z)] \equiv (Z-1) T_1(Z)$$
$$T_1(Z) = [-Z^{q-2}-2Z^{q-3}...-(q-3)Z^2-(q-2)Z+1-q]/q. \tag{3.32}$$

Since $T_1(Z)$ is defined modulo $P(Z)$, premultiplication by $T_1(Z)$ can be done prior to Chinese remainder reconstruction and combined with the precomputation of the transform of the constant sequence. Similarly premultiplication by $1/q$ can be combined with the computation of the scalar convolution $Y_{2,l}$ in (3.26) so that the Chinese remainder reconstruction given in (3.22) reduces to

$$Y_l(Z) \equiv (Z-1) Y_{1,l}(Z) + (Z^{q-1}+Z^{q-2}+...+1) Y_{2,l} \text{ modulo } (Z^q-1). \tag{3.33}$$

Under these conditions, the convolution of $q \times q$ points is computed as shown in Fig. 3.1. It can be seen that the only multiplications required to calculate $y_{u,l}$ correspond to the computation of one convolution of q points and of q polynomial products modulo $P(Z)$. This means that, if the convolution and polynomial products are evaluated with the minimum multiplication algorithms defined by Theorems 3.2 and 3.3, the convolution of $q \times q$ points with q prime is computed with only $(2q^2-q-2)$ multiplications. It can indeed be shown that this is the theoretical minimum number of multiplications for a convolution of $q \times q$ points, with q prime [3.13].

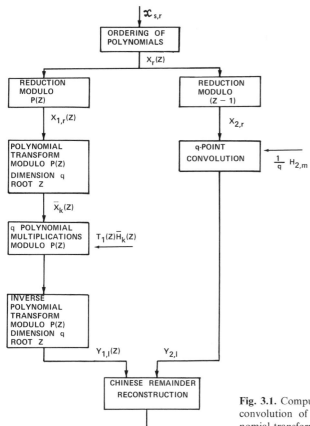

Fig. 3.1. Computation of a two-dimensional convolution of dimension $N \times N$ by polynomial transforms. N prime

3.2.2 Composite Polynomial Transforms

In order to simplify the presentation, we have, up to now, restricted our discussion to convolutions of $N \times N$ terms, with $N = q$, q prime. In practice, this polynomial transform approach is not restricted to N prime. It is clear, from (3.28–31) that any polynomial transform having a root aZ^b and defined modulo a polynomial $P_{e_i}(Z)$ will have the circular convolution property provided

$$(aZ^b)^N \equiv 1 \text{ modulo } P_{e_i}(Z) \tag{3.34}$$

$$S \equiv \begin{cases} 0 \text{ modulo } P_{e_i}(Z) & \text{for } p \not\equiv 0 \text{ modulo } N \\ N \text{ modulo } P_{e_i}(Z) & \text{for } p \equiv 0 \text{ modulo } N \end{cases} \tag{3.35}$$

with $S = \sum_{k=0}^{N-1} (aZ^d)^{pk}$. Since polynomial transforms are used here for computing two-dimensional convolutions, $P_{e_i}(Z)$ is a cyclotomic polynomial, factor of

$(Z^N - 1)$, with

$$Z^N - 1 = \prod_{i=1}^{d} P_{e_i}(Z) \tag{3.36}$$

where the degree of each polynomial $P_{e_i}(Z)$ is $\varphi(e_i)$, the Euler's totient function [3.7]. We show first that there is always a polynomial transform, of dimension N and root Z, having the circular convolution property when defined modulo $P_{e_d}(Z)$, the largest cyclotomic polynomial, factor of $(Z^N - 1)$. This can be seen by noting that, since $Z^N \equiv 1$ modulo $(Z^N - 1)$ and $P_{e_d}(Z)$ is a factor of $(Z^N - 1)$, $Z^N \equiv 1$ modulo $P_{e_d}(Z)$, thus satisfying (3.34). Condition (3.35) is verified for $p \equiv 0$ modulo N because $S \equiv \sum_{k=0}^{N-1} Z^{pk} = N$. For $p \not\equiv 0$ modulo N and relatively prime with N, $S \equiv \sum_{k=0}^{N-1} Z^k$ modulo $(Z^N - 1)$, which implies that $S = \prod_{i=2}^{d} P_{e_i}(Z) \equiv 0$ modulo $P_{e_d}(Z)$. For $p \not\equiv 0$ modulo N and non prime with N, p can always be considered, without loss of generality as a factor of N. Then, $S \equiv p \sum_{k=0}^{N/p-1} Z^{pk} = p(Z^N - 1)/(Z^p - 1)$. The largest polynomial factor of $(Z^N - 1)$ is $P_{e_d}(Z)$ of degree $\varphi(N)$. The largest polynomial factor of $(Z^p - 1)$ is a cyclotomic polynomial $Q(Z)$ of degree $\varphi(p)$. $Q(Z)$ is different from $P_{e_d}(Z)$ since $\varphi(p) < \varphi(N)$ and it cannot be a factor of $P_{e_d}(Z)$ since $P_{e_d}(Z)$ is irreducible. Thus, $Z^p - 1 \not\equiv 0$ modulo $P_{e_d}(Z)$ and $S \equiv 0$ modulo $P_{e_d}(Z)$, which completes the verification of (3.35).

Under these conditions, the convolution $y_{u,l}$ of $N \times N$ points is computed by arranging the input sequence into N polynomials of N terms which are reduced modulo $P_{e_d}(Z)$ and modulo $\prod_{i=1}^{d-1} P_{e_i}(Z)$. The output samples $y_{u,l}$ are derived by a Chinese remainder reconstruction from a polynomial convolution $Y_{1,l}(Z)$ modulo $P_{e_d}(Z)$ and from a polynomial convolution $Y_{2,l}(Z)$ modulo $\prod_{i=1}^{d-1} P_{e_i}(Z)$. $Y_{1,l}(Z)$ is computed by polynomial transforms of dimension N. Since the degree of $P_{e_d}(Z)$ is $\varphi(N)$, the degree of $\prod_{i=1}^{d-1} P_{e_i}(Z)$ is $N - \varphi(N)$ and $Y_{2,l}(Z)$ can be viewed as a generalized scalar convolution of $N \times (N - \varphi(N))$ points. The same method can then be used recursively to compute $Y_{2,l}(Z)$ by polynomial transforms.

It should also be noted that when a polynomial transform of root aZ^b and of dimension N_1, with N_1 odd, has the convolution property, it is always possible to increase the length of this polynomial transform to $N_1 N_2$, with $N_2 = 2^t$. In this case, the root of the new transform is aWZ^b, where $W = \exp(-j2\pi/N_2), j = \sqrt{-1}$. This property, which is a direct consequence of the fact that W is a root of unity of order N_2 and aZ^b is a root of unity of order N_1, with N_1, N_2 relatively prime is particularly useful to compute convolutions of dimensions $2N_1 \times 2N_1$ and $4N_1 \times 4N_1$, because in these cases $W = -1$ or $-j$ and the polynomial transforms are still computed without multiplications.

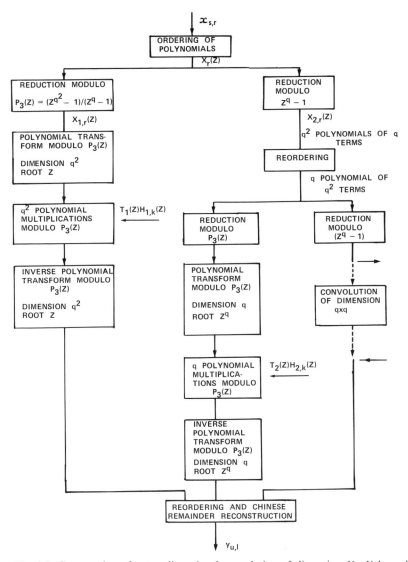

Fig. 3.2. Computation of a two-dimensional convolution of dimension $N \times N$ by polynomial transforms. $N = q^2$, q prime

For N composite, the main cases of interest correspond to N being a power of a prime. We illustrate in Fig. 3.2 the computation of the convolution of dimension $q^2 \times q^2$, with q odd prime, by polynomial transforms. In this example, the convolution of $q^2 \times q^2$ points is mapped without multiplications into $(q^2 + q)$ polynomial products modulo $P_3(Z) = (Z^{q^2} - 1)/(Z^q - 1) = Z^{q(q-1)} + Z^{q(q-2)} + \ldots + 1$ and into a convolution of

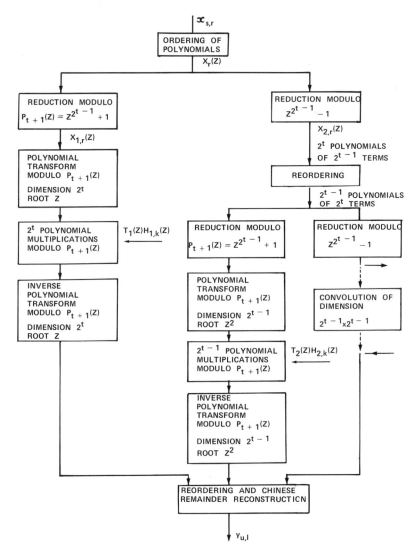

Fig. 3.3. Computation of a two-dimensional convolution of dimension $N \times N$ by polynomial transforms. $N = 2^t$

$q \times q$ points. This last convolution can itself be mapped without multiplications by the approach of Fig. 3.1 into q polynomial products modulo $P_2(Z) = Z^{q-1} + Z^{q-2} + \ldots + 1$ and one convolution of q points. We show also, in Fig. 3.3, the first stage of the calculation of a convolution of dimension $2^t \times 2^t$ by polynomial transforms. In this case, the computation is performed in $(t-1)$ stages with polynomial transforms defined modulo $P_{t+1}(Z) = Z^{2^{t-1}} + 1$,

Table 3.1. Polynomial transforms for the computation of convolutions. q, q_1, q_2 primes

Transform ring P(Z)	Transform		Dimension of convolutions NxN	No. of additions for reductions, polynomial transforms and Chinese remainder reconstruction	Polynomial products and convolutions
	Length	Root			
$(Z^q-1)/(Z-1)$	q	Z	qxq	$2(q^3+q^2-5q+4)$	q products P(Z) 1 convolution q
"	2q	-Z	2qxq	$4(q^3+2q^2-6q+4)$	2q products P(Z) 1 convolution 2q
$(Z^{2q}-1)/(Z^2-1)$	2q	$-Z^{q+1}$	2qx2q	$8(q^3+2q^2-6q+4)$	2q products P(Z) 1 convolution 2x2q
$(Z^{q^2}-1)/(Z^q-1)$	q	Z^q	qxq^2	$2(q^4+2q^3-4q^2$ $-q+4)$	q products P(Z) q products $(Z^q-1)/(Z-1)$ 1 convolution q
"	q^2	Z	q^2xq^2	$2(2q^5+q^4-5q^3$ $+q^2+6)$	q(q+1) products P(Z) q products $(Z^q-1)/(Z-1)$ 1 convolution q
$(Z^{2q^2}-1)/(Z^{2q}-1)$	$2q^2$	$-Z^{q^2+1}$	$2q^2x2q^2$	$8(2q^5+2q^4-6q^3$ $-q^2+5q+2)$	2q(q+1) products P(Z) 2q products $(Z^{2q}-1)/(Z^2-1)$ 1 convolution 2x2q
$(Z^{q_1q_2}-1)/(Z^{q_2}-1)$	q_1	Z^{q_2}	$q_1xq_1q_2$	$2q_2(q^3{}_1+q^2{}_1$ $-5q_1+4)$	q_1 products P(Z) 1 convolution q_1q_2
$Z^{2^{t-1}}+1$	2^t	Z	2^tx2^t	$2^{2t-1}(3t+5)$	3.2^{t-1} products P(Z) 1 convolution $2^{t-1}x2^{t-1}$

$P_t(Z)=Z^{2^{t-2}}+1, \ldots, P_3(Z)=Z^2+1$. These polynomial transforms are particularly interesting because their dimensions are powers of two so that they can be computed with a reduced number of additions by a radix-2 FFT-type algorithm.

We summarize in Table 3.1 the main existence conditions for polynomial transforms that can be computed without multiplications and which correspond to two-dimensional convolutions such that both dimensions have a common factor.

3.2.3 Computation of Polynomial Transforms and Reductions

The evaluation of two-dimensional convolutions by polynomial transforms involves the calculation of reductions, Chinese remainder reconstructions and polynomial transforms. We specify here the number of additions required for these various operations corresponding to the main cases of interest summarized in Table 3.1.

When $N=q$, q prime, the input sequences must be reduced modulo $(Z-1)$ and modulo $P(Z)=Z^{q-1}+Z^{q-2}+...+1$. Each of these operations is done very simply, with $q(q-1)$ additions, by summing all terms of the input sequence for the reduction modulo $(Z-1)$, as shown in (3.27) and by noting that the reduction modulo $P(Z)$ yields N polynomials $H_{1,m}(Z)$ defined by

$$H_{1,m}(Z)= \sum_{n=0}^{q-2} (h_{n,m}-h_{q-1,m}) Z^n \quad m=0,...,q-1. \tag{3.37}$$

The Chinese remainder reconstruction is done with $2q(q-1)$ additions by using (3.33). Since $Y_{1,l}(Z)$ is a polynomial of $(q-1)$ terms, $Y_{1,l}(Z)$ can be defined by

$$Y_{1,l}(Z)= \sum_{u=0}^{q-2} y_{1,u,l} Z^u. \tag{3.38}$$

Thus, (3.33) becomes

$$Y_l(Z)= Y_{2,l}-y_{1,0,l}+ \sum_{u=1}^{q-2} (y_{1,u-1,l}-y_{1,u,l}+ Y_{2,l}) Z^u$$

$$+(y_{1,q-2,l}+ Y_{2,l}) Z^{q-1}, \quad l=0,...,q-1 \tag{3.39}$$

for $N=q^2$, q prime or for $N=2^t$, the reductions and Chinese remainder reconstruction are implemented in a similar fashion and we give in Table 3.2, column 3, the corresponding number of additions.

For $N=q$, q prime, the polynomial transform $\bar{H}_k(Z)$ defined by (3.28) is computed with

$$\bar{H}_k(Z)\equiv H_{1,0}(Z)+ \sum_{m=1}^{q-1} H_{1,m}(Z) Z^{mk} \text{ modulo } P(Z), \quad k=0,...,q-2 \tag{3.40}$$

$$\bar{H}_{q-1}(Z)\equiv H_{1,0}(Z)+ \sum_{m=1}^{q-1} H_{1,m}(Z) Z^{m(q-1)} \text{ modulo } P(Z) \tag{3.41}$$

where $H_{1,m}(Z)$ are polynomials of $(q-1)$ terms. Let $R_k(Z)\equiv \sum_{m=1}^{q-1} H_{1,m}(Z) Z^{mk}$. Since $R_0(Z)= \sum_{m=1}^{q-1} H_{1,m}(Z)$, $R_0(Z)$ is computed with $(q-2)(q-1)$ additions. For $k\neq 0$, $R_k(Z)$ is the sum of $(q-1)$ polynomials, each multiplied by a power of Z. $R_k(Z)$ is first computed modulo (Z^q-1). Since $Z^q\equiv 1$, each polynomial $H_{1,m}(Z)$ becomes, after multiplication by a power of Z, a polynomial of q words where one of the words is zero and the calculation of $R_k(Z)$ modulo (Z^q-1) requires (q^2-3q+1) additions. $R_k(Z)$ is then reduced modulo $P(Z)$ with $(q-1)$ ad-

Table 3.2. Number of additions for the computation of reductions, Chinese remainder operations and polynomial transforms. q, q_1, q_2 odd primes

Transform size	Ring	Number of additions for reductions and Chinese remainder operations	Number of additions for polynomial transforms
q	$(Z^q-1)/(Z-1)$	$4q(q-1)$	q^3-q^2-3q+4
$2q$	$(Z^q-1)/(Z-1)$	$8q(q-1)$	$2(q^3-4q+4)$
$2q$	$(Z^{2q}-1)/(Z^2-1)$	$16q(q-1)$	$4(q^3-4q+4)$
q	$(Z^{q^2}-1)/(Z^q-1)$	$4q^2(q-1)$	$q(q^3-q^2-3q+4)$
q^2	$(Z^{q^2}-1)/(Z^q-1)$	$4q^3(q-1)$	$2q^5-2q^4-5q^3+5q^2+q+2$
$2q$	$(Z^{2q^2}-1)/(Z^{2q}-1)$	$16q^2(q-1)$	$4q(q^3-4q+4)$
$2q^2$	$(Z^{2q^2}-1)/(Z^{2q}-1)$	$16q^3(q-1)$	$4(2q^5-q^4-6q^3+5q^2+q+2)$
2^t	$Z^{2^{t-1}}+1$	2^{2t+1}	$t2^{2t-1}$

ditions. Thus, computing each $R_k(Z)$ for $k \neq 0$ requires $(q^2 - 2q)$ additions. Since

$$\sum_{k=0}^{q-2} Z^{mk} \equiv -Z^{m(q-1)} \text{ modulo } P(Z), \text{ for } m \neq 0,$$

$$\sum_{k=0}^{q-2} \sum_{m=1}^{q-1} H_{1,m}(Z) Z^{mk} \equiv - \sum_{m=1}^{q-1} H_{1,m}(Z) Z^{m(q-1)} \quad \text{modulo } P(Z) \tag{3.42}$$

and $\sum_{m=1}^{q-1} H_{1,m}(Z) Z^{m(q-1)}$ is calculated with $(q-1)(q-2)$ additions by

$$\sum_{m=1}^{q-1} H_{1,m}(Z) Z^{m(q-1)} \equiv - \sum_{k=0}^{q-2} R_k(Z) \quad \text{modulo } P(Z). \tag{3.43}$$

Finally, $\bar{H}_k(Z)$ is obtained by adding $H_{1,0}(Z)$ to $R_k(Z)$ and to $\sum_{m=1}^{q-1} H_{1,m}(Z) Z^{m(q-1)}$ with $q(q-1)$ additions. Thus, a total of $(q^3 - q^2 - 3q + 4)$ additions is required to compute a polynomial transform of q terms, with q prime.

For N composite, the polynomial transforms are computed with a reduced number of additions by using an FFT-type algorithm. Assuming, for instance $N = N_1 N_2$, the polynomial transform $\bar{H}_k(Z)$ of dimension N is computed with N_1 transforms of dimension N_2, N_2 transforms of dimension N_1 and twiddle factors $Z^{m_2 k_2}$:

$$\bar{H}_{N_2 k_1 + k_2}(Z) \equiv \sum_{m_2=0}^{N_1-1} Z^{N_2 m_2 k_1} Z^{m_2 k_2} \sum_{m_1=0}^{N_2-1} H_{1, N_1 m_1 + m_2}(Z) Z^{N_1 m_1 k_2} \tag{3.44}$$

$$k_1 = 0, ..., N_1 - 1$$

$$k_2 = 0, ..., N_2 - 1.$$

We summarize in Table 3.2, column 4, the number of additions for various polynomial transforms. We also give in Table 3.1, column 4, the total number of additions for reductions, polynomial transforms and Chinese remainder reconstruction for various two-dimensional convolutions evaluated by polynomial transforms.

3.2.4 Computation of Polynomial Products and One-Dimensional Convolutions

We have seen in the preceding sections that two-dimensional convolutions can be evaluated efficiently by mapping the two-dimensional convolutions into one-dimensional polynomial products and convolutions by means of polynomial transforms. When the polynomial transforms are properly selected, the mapping is done without multiplications and with a limited number of additions. Thus, when two-dimensional convolutions are evaluated by polynomial transforms, the processing load is strongly dependent upon the efficiency of the algorithms used for the calculation of polynomial products and one-dimensional convolutions.

A first approach for evaluating one-dimensional convolutions and polynomial products consists in using DFTs or NTTs. Since these transforms have the convolution property, they can be used directly to compute the one-dimensional convolutions. These transforms can also be used to compute polynomial products modulo $P_{e_i}(Z)$ by noticing that, since $P_{e_i}(Z)$ defined by (3.36) is a factor of $(Z^N - 1)$, all computations can be carried out modulo $(Z^N - 1)$ with a final reduction modulo $P_{e_i}(Z)$. With this method, the calculation of a polynomial product modulo $P_{e_i}(Z)$ is replaced by that of a polynomial product modulo $(Z^N - 1)$, which is a convolution of dimension N. Hence, calculations of polynomial products modulo $P_{e_i}(Z)$, where $P_{e_i}(Z)$ is a factor of $(Z^N - 1)$ can be done by using DFTs or NTTs of dimension N. This method can be illustrated by the following simple example: assuming the two input polynomials $X(Z) = x_0 + x_1 Z$ and $H(Z) = h_0 + h_1 Z$ are to be multiplied modulo $P(Z) = (Z^3 - 1)/(Z - 1) = Z^2 + Z + 1$, the computation can be performed by first calculating the 3-points circular convolution $Y(Z)$ of the two input sequences with

$$Y(Z) = y_0 + y_1 Z + y_2 Z^2 \tag{3.45}$$

$$\begin{cases} y_0 = h_0 x_0 \\ y_1 = h_0 x_1 + h_1 x_0 \\ y_2 = h_1 x_1. \end{cases} \tag{3.46}$$

The polynomial product $Y_1(Z)$, defined modulo $P(Z)$ is then given by $Y_1(Z) \equiv \{[H(Z)X(Z)] \text{ modulo } (Z^N - 1)\} \text{ modulo } P(Z)$ which corresponds to a simple reduction of $Y(Z)$ modulo $P(Z)$. Since $Z^2 \equiv -Z - 1$, this reduction is achieved by simply subtracting y_2 to y_0 and y_1, with

$$Y_1(Z) = y_{1,0} + y_{1,1} Z \equiv Y(Z) \text{ modulo } P(Z) \tag{3.47}$$

$$\begin{cases} y_{1,0} = h_0 x_0 - h_1 x_1 \\ y_{1,1} = h_0 x_1 + h_1 x_0 - h_1 x_1. \end{cases} \tag{3.48}$$

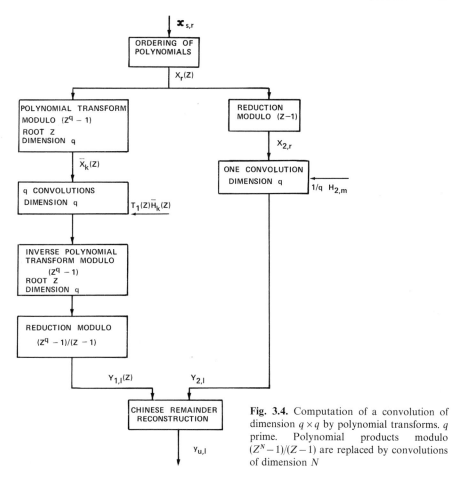

Fig. 3.4. Computation of a convolution of dimension $q \times q$ by polynomial transforms. q prime. Polynomial products modulo $(Z^N - 1)/(Z - 1)$ are replaced by convolutions of dimension N

This method is shown in Fig. 3.4 for a convolution of dimension $q \times q$, with q prime. In this case, the two-dimensional convolution is mapped into $(q+1)$ convolutions of dimension q instead of one convolution of dimension q plus q polynomial products modulo $(Z^q - 1)/(Z - 1)$, with the method given in Fig. 3.1.

Thus, the one-dimensional convolutions and polynomial products can be computed by DFTs or NTTs. The problems related to the use of these transforms such as roundoff errors for DFTs or modular arithmetic for NTTs are then limited to only a part of the total computation process. We have seen however in Sect. 3.1 that the methods based on interpolation and on the Chinese remainder theorem yield more efficient algorithms than the DFTs or NTTs for short convolutions and polynomial products. It is therefore often advantageous to use such algorithms in combination with polynomial transforms. We shall see that these algorithms cannot be derived systematically from a general method as is the case, for instance, for FFT algorithms. Thus, each

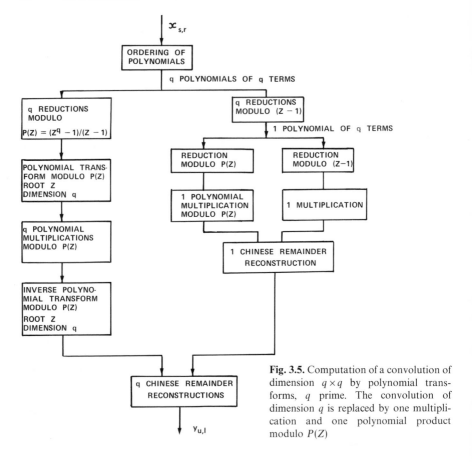

Fig. 3.5. Computation of a convolution of dimension $q \times q$ by polynomial transforms, q prime. The convolution of dimension q is replaced by one multiplication and one polynomial product modulo $P(Z)$

short convolution or polynomial product algorithm used in a given application must be specifically programmed and it is desirable to use only a limited number of different algorithms in order to reduce program sizes. One way to do it consists in computing the short convolutions as polynomial products by using the Chinese remainder theorem. In the case of a two-dimensional convolution of dimension $q \times q$, with q prime, the polynomial transform approach requires the computation of q polynomial products modulo $P(Z) = (Z^q - 1)/(Z - 1)$ and one circular convolution of dimension q. However, the computation can be done with only one polynomial product algorithm modulo $P(Z)$ if the circular convolution of dimension q is computed as one multiplication and one polynomial product modulo $P(Z)$ by using the Chinese remainder theorem. In this case, the convolution of dimension $q \times q$ is computed with $(q + 1)$ polynomial products modulo $P(Z)$ and one multiplication instead of q polynomial products modulo $P(Z)$ and one convolution of dimension q. The process is illustrated in Fig. 3.5.

In the following, we will assume that all two-dimensional convolutions are computed in a similar manner so that we need to be concerned only with polynomial product algorithms. Moreover, we shall see that in most practical cases, two-dimensional convolutions can be computed with a limited set of polynomial product algorithms, mainly those defined modulo $P(Z)=(Z^q-1)/(Z-1)$, with $q=3, 5, 7$, and modulo $P(Z)=(Z^9-1)/(Z^3-1)$ or $(Z^{2^t}+1)$. Since these polynomials $P(Z)$ are irreducible, they always can be computed by interpolation, according to Theorem 3.2, with $D-1$ multiplications, where D is the degree of $P(Z)$. Using this method for multiplications modulo (Z^2+1) and $(Z^3-1)/(Z-1)$ yields algorithms with 3 multiplications and 3 additions as shown in Sects. 3.5.1, 2.

For longer polynomial products, however, this approach which gives a minimum number of multiplications, yields an excessive number of additions. This is due to the fact that the minimum multiplication method requires $2D-1$ distinct polynomials $(Z-a_i)$. The four simplest interpolation polynomials are $Z, 1/Z, Z-1$, and $Z+1$. Thus, when the degree D of $P(Z)$ is larger than 2, one must use integers a_i different from 0 or ± 1, and the corresponding reductions modulo $(Z-a_i)$ and Chinese remainder reconstructions will involve multiplications by powers of a_i which have to be implemented either with scalar multiplications or by a large number of successive additions. This leads to design compromise algorithms in which a reasonable balance is kept between the number of additions and the number of multiplications. Although, there are no general methods for designing such algorithms, the few following principles are useful.

A first technique for reducing the number of additions at the expense of a slightly increased number of multiplications consists in using complex interpolation polynomials. If we use, for instance $Z^2+1=(Z+j)(Z-j), j=\sqrt{-1}$, we need 3 real multiplications instead of 2 for a polynomial having real integer roots, but reductions and Chinese remainder reconstruction remain simple since these operations require only simple additions and multiplications by ± 1 or $\pm j$. Another approach consists in converting one-dimensional polynomial products into multidimensional polynomial products. For instance, a multiplication modulo $P(Z)=(Z^9-1)/(Z^3-1)=Z^6+Z^3+1$, can be replaced by a multiplication modulo $(Z^3-Z_1), (Z_1^2+Z_1+1)$, since substituting $Z_1\equiv Z^3$ into $Z_1^2+Z_1+1$ yields $P(Z)$. With this method, the input polynomials $X(Z)=\sum_{n=0}^{5}x_nZ^n$ and $H(Z)=\sum_{m=0}^{5}h_mZ^m$ are first reduced modulo (Z^3-Z_1) by substituting $Z^3\equiv Z_1$ into $X(Z)$ and $H(Z)$. Thus, $X(Z)$ becomes

$$X(Z)=X_0(Z)+ZX_1(Z)+Z^2X_2(Z) \tag{3.49}$$

$$\begin{cases} X_0(Z)=x_0+x_3Z_1 \\ X_1(Z)=x_1+x_4Z_1 \\ X_2(Z)=x_2+x_5Z_1 \end{cases} \tag{3.50}$$

with similar relations for $H(Z)$. The polynomial product modulo $(Z^3 - Z_1)$ can then be computed by interpolation with 5 multiplications modulo $(Z_1^2 + Z_1 + 1)$. Since each of these polynomial multiplications requires 3 scalar multiplications as shown in Sect. 3.5.2, the polynomial product modulo $(Z^9 - 1)/(Z^3 - 1)$ is implemented with 15 multiplications. The detailed algorithm is given in Sect. 3.5.5.

A similar technique can be used for polynomial products modulo $(Z^{2^t} + 1)$. A polynomial multiplication modulo $(Z^4 + 1)$ is, for instance, implemented modulo $(Z^2 - Z_1), (Z_1^2 + 1)$ with 9 multiplications and 15 additions (Sect. 3.5.3), while the conventional interpolation method would have required 7 multiplications and 41 additions.

All the algorithms designed with these techniques can be viewed as a computation by rectangular transforms. Assuming a polynomial product algorithm modulo a polynomial $P(Z)$ of degree N and requiring M multiplications, the general structure of the algorithm is given by

$$H = Eh \tag{3.51}$$

$$X = Fx \tag{3.52}$$

$$Y = H \times X \tag{3.53}$$

$$y = GY \tag{3.54}$$

where h and x are the column vectors of the input sequences $\{h_m\}$ and $\{x_n\}$, E and F are the input matrices of size $M \times N$, $\times$ denotes the element by element product, G is the output matrix of size $N \times M$ and y is the column vector of the output sequence $\{y_1\}$. Eh, Fx, and GY are computed without multiplications, while $H \times X$ is evaluated with M multiplications. When the algorithm is designed by an interpolation method, the matrices E and F correspond to the various reductions modulo $(Z - a_i)$ while the matrix G correspond to the Chinese remainder reconstruction. Since Chinese remainder reconstruction can be seen as the inverse operation of the reductions, it is roughly equivalent in complexity to two reductions so that the matrix G is usually about twice as complex as matrices E and F. In most practical filtering applications, one of the input sequence, $\{h_m\}$, is constant so that H is precomputed and stored. Thus, the number of additions for the algorithm is primarily dependent upon the complexity of F and G matrices but not of that of E matrix. Under these conditions, it would be highly desirable to permute E and G matrices in order to reduce the number of additions. One way to do it is to note that, from (3.50–54), y_1 is given by

$$y_l = \sum_{n=0}^{N-1} \sum_{m=0}^{N-1} h_m x_n \sum_{k=0}^{M-1} e_{m,k} f_{n,k} g_{k,l} \tag{3.55}$$

where $e_{m,k}$, $f_{n,k}$ and $g_{k,l}$ are, respectively, the elements of matrices E, F, and G. For an algorithm modulo $P(Z) = Z^N - 1$, corresponding to a circular convolution, we find the usual condition for the circular convolution property with

$$S = \sum_{k=0}^{M-1} e_{m,k} f_{n,k} g_{k,l} \begin{matrix} =1 \\ =0 \end{matrix} \quad \begin{matrix} \text{if} \\ \text{if} \end{matrix} \quad \begin{matrix} m+n-l \equiv 0 \text{ modulo } N \\ m+n-l \not\equiv 0 \text{ modulo } N. \end{matrix} \tag{3.56}$$

Similarly, polynomial product algorithms modulo $(Z^N + 1)$ or $(Z^N - Z_1)$ correspond, respectively, to the conditions

$$\begin{cases} S = 1 & \text{if} \quad m+n-l = 0 \\ S = -1 \text{ or } Z_1 & \text{if} \quad m+n-l = N \\ S = 0 & \text{if} \quad m+n-l \not\equiv 0 \text{ modulo } N. \end{cases} \tag{3.57}$$

We now replace the matrices E and G by matrices E^1 and G^1 such that their elements are, respectively, $g_{k,N-m}$ and $e_{N-l,k}$. Under these conditions, S becomes S^1 and (3.56) obviously implies

$$S^1 = \sum_{k=0}^{M-1} e_{N-l,k} f_{n,k} g_{k,N-m} \begin{matrix} =1 \\ =0 \end{matrix} \quad \begin{matrix} \text{if} \\ \text{if} \end{matrix} \quad \begin{matrix} m+n-l \equiv 0 \text{ modulo } N \\ m+n-l \not\equiv 0 \text{ modulo } N. \end{matrix} \tag{3.58}$$

Thus, as pointed out in [3.14], the convolution property still holds if the matrices E and G are exchanged with simple transposition and rearrangement

Table 3.3. Number of operations for polynomial product algorithms

Ring P(Z)	Degree of P(Z)	Number of multiplications	Number of additions
$Z^2 + 1$	2	3	3
$(Z^3 - 1)/(Z - 1)$	2	3	3
$Z^4 + 1$	4	9	15
$Z^4 + 1$	4	7	41
$(Z^5 - 1)/(Z - 1)$	4	9	16
$(Z^5 - 1)/(Z - 1)$	4	7	46
$(Z^7 - 1)/(Z - 1)$	6	15	53
$(Z^9 - 1)/(Z^3 - 1)$	6	15	39
$Z^8 + 1$	8	21	77
$(Z^2_1 + Z_1 + 1)(Z^5_2 - 1)/(Z_2 - 1)$	8	21	83
$(Z^4_1 + 1)(Z^2_2 + Z_2 + 1)$	8	21	76
$(Z^2_1 + 1)(Z^5_2 - 1)/(Z - 1)$	8	21	76
$Z^{16} + 1$	16	63	205
$(Z^{27} - 1)/(Z^9 - 1)$	18	75	267
$Z^{32} + 1$	32	147	739
$Z^{64} + 1$	64	315	1891

Table 3.4. Number of operations for two-dimensional convolutions computed by polynomial transforms

Convolution size	Number of multiplications	Number of additions	Multiplications per point	Additions per point
3x3 (9)	13	70	1.44	7.78
4x4 (16)	22	122	1.37	7.62
5x5 (25)	55	369	2.20	14.76
6x6 (36)	52	424	1.44	11.78
7x7 (49)	121	1163	2.47	23.73
8x8 (64)	130	750	2.03	11.72
9x9 (81)	193	1382	2.38	17.06
10x10 (100)	220	1876	2.20	18.76
14x14 (196)	484	5436	2.47	27.73
16x16 (256)	634	4774	2.48	18.65
18x18 (324)	772	6576	2.38	20.30
24x24 (576)	1402	12954	2.43	22.49
27x27 (729)	2893	21266	3.97	29.17
32x32 (1024)	3658	24854	3.57	24.27
64x64 (4096)	17770	142902	4.34	34.89
128x128 (16384)	78250	718966	4.78	43.88

of lines and columns. The same general approach can be used for polynomial products modulo $(Z^N + 1)$ or $(Z^N - Z_1)$. However, in these cases, the conditions $S = 1$ and $S = -1$ or Z_1 in (3.57) are inversed for m or $l = 0$. Thus, the elements of E^1 and G^1 must be $g_{k,N-m}$ for $m \neq 0$, $-g_{k,N-m}$ or $(1/Z_1)g_{k,N-m}$ for $m = 0$ and $e_{N-l,k}$ for $l \neq 0$, $-e_{N-l,k}$ or $Z_1 e_{N-l,k}$ for $l = 0$. This approach has been used for the polynomial product algorithm modulo $(Z^9 - 1)/(Z^3 - 1)$ given in Sect. 3.5.5.

We give in Appendix 3.5 the detailed algorithms for the most frequently used polynomial products and in Table 3.3 the corresponding number of operations. The number of arithmetic operations for two-dimensional convolutions computed with these algorithms is given in Table 3.4. It should be noted that we have optimized our algorithms by weighting heavily in favor of reducing the number of multiplications. When the calculations are done in a computer in which the execution times for multiplications are about the same as for additions, it is preferable to use other polynomial product algorithms in which the number of additions is further reduced at the expense of an increased number of multiplications.

3.2.5 Nesting Algorithms

A systematic application of the techniques discussed in the preceding section allows one to design large polynomial product algorithms using a relatively low number of additions and multiplications and therefore to compute large two-dimensional convolutions by composite polynomial transforms. An alternative to this approach is to construct large two-dimensional convolutions from a limited set of short two-dimensional convolutions with a technique similar to that used for one-dimensional convolutions [3.2]. In this method, a two-dimensional convolution $y_{u,l}$ of dimension $N_1N_2 \times N_1N_2$, where N_1 and N_2 are relatively prime, is converted into a four-dimensional convolution of dimension $(N_1 \times N_1) \times (N_2 \times N_2)$ by a simple index mapping. $y_{u,l}$ is given by

$$y_{u,l} = \sum_{m=0}^{N_1N_2-1} \sum_{n=0}^{N_1N_2-1} h_{n,m} x_{u-n,l-m}. \tag{3.59}$$

Since N_1 and N_2 are relatively prime and the indices l, m, n, and u are defined modulo N_1N_2, a direct consequence of the Chinese remainder theorem [3.7] is that l, m, n, and u can be mapped into two sets of indices l_1, m_1, n_1, u_1 and l_2, m_2, n_2, u_2 with

$$\begin{cases} l \equiv N_1 l_2 + N_2 l_1 \text{ modulo } N_1 N_2 \\ m \equiv N_1 m_2 + N_2 m_1 \text{ modulo } N_1 N_2 \\ n \equiv N_1 n_2 + N_2 n_1 \text{ modulo } N_1 N_2, \quad l_1, m_1, n_1, u_1 = 0, ..., N_1 - 1 \\ u \equiv N_1 u_2 + N_2 u_1 \text{ modulo } N_1 N_2, \quad l_2, m_2, n_2, u_2 = 0, ..., N_2 - 1. \end{cases} \tag{3.60}$$

Hence, $y_{u,l}$ becomes a four-dimensional convolution of dimension $(N_1 \times N_1) \times (N_2 \times N_2)$

$$y_{N_1 u_2 + N_2 u_1, N_1 l_2 + N_2 l_1} = \sum_{m_1=0}^{N_1-1} \sum_{n_1=0}^{N_1-1} \sum_{m_2=0}^{N_2-1} \sum_{n_2=0}^{N_2-1} h_{N_1 n_2 + N_2 n_1, N_1 m_2 + N_2 m_1}$$
$$\cdot x_{N_1(u_2-n_2)+N_2(u_1-n_1), N_1(l_2-m_2)+N_2(l_1-m_1)}. \tag{3.61}$$

This four-dimensional convolution can be viewed as a two-dimensional convolution of dimension $N_1 \times N_1$ where all scalars are replaced by two-dimensional polynomials of dimension $N_2 \times N_2$. More precisely, by organizing the input and output sequences into two-dimensional polynomials, we have

$$H_{n_1,m_1}(Z_1, Z_2) = \sum_{m_2=0}^{N_2-1} \sum_{n_2=0}^{N_2-1} h_{N_1 n_2 + N_2 n_1, N_1 m_2 + N_2 m_1} Z_1^{m_2} Z_2^{n_2} \tag{3.62}$$

$$X_{r_1,s_1}(Z_1, Z_2) = \sum_{r_2=0}^{N_2-1} \sum_{s_2=0}^{N_2-1} x_{N_1 r_2 + N_2 r_1, N_1 s_2 + N_2 s_1} Z_1^{r_2} Z_2^{s_2} \quad r_1, s_1 = 0, ..., N_1 - 1 \tag{3.63}$$

$$Y_{u_1,l_1}(Z_1, Z_2) = \sum_{u_2=0}^{N_2-1} \sum_{l_2=0}^{N_2-1} y_{N_1 u_2 + N_2 u_1, N_1 l_2 + N_2 l_1} Z_1^{u_2} Z_2^{l_2} \tag{3.64}$$

$$Y_{u_1,l_1}(Z_1, Z_2) \equiv \sum_{m_1=0}^{N_1-1} \sum_{n_1=0}^{N_1-1} H_{n_1,m_1}(Z_1, Z_2) X_{u_1-n_1, l_1-m_1}(Z_1, Z_2)$$
$$\cdot \text{modulo } (Z_1^{N_2} - 1), (Z_2^{N_2} - 1). \tag{3.65}$$

Table 3.5. Number of operations for two-dimensional convolutions computed by polynomial transforms and nesting

Convolution size	Number of multiplications	Number of additions	Multiplications per point	Additions per point
12x12 (144)	286	2638	1.99	18.32
20x20 (400)	946	14030	2.36	35.07
30x30 (900)	2236	35404	2.48	39.34
36x36 (1296)	4246	40286	3.28	31.08
40x40 (1600)	4558	80802	2.85	50.50
60x60 (3600)	12298	195414	3.42	54.28
72x72 (5184)	20458	232514	3.95	44.85
80x80 (6400)	27262	345826	4.26	54.03
120x120 (14400)	59254	1081756	4.11	75.12

Each polynomial multiplication $H_{n_1,m_1}(Z_1, Z_2) X_{u_1-n_1, l_1-m_1}(Z_1, Z_2)$ modulo $(Z_1^{N_2} - 1), (Z_2^{N_2} - 1)$ corresponds to a convolution of dimension $N_2 \times N_2$ which is computed with M_2 scalar multiplications and A_2 scalar additions. In the convolution of dimension $N_1 \times N_1$, all scalars are replaced by polynomials of dimension $N_2 \times N_2$. Thus, if M_1 and A_1 are the number of multiplications and additions required to compute a scalar convolution of dimension $N_1 \times N_1$, the number of multiplications M and additions A required to evaluate the two-dimensional convolution of dimension $N_1 N_2 \times N_1 N_2$ reduces to

$$M = M_1 M_2 \tag{3.66}$$

$$A = N_2^2 A_1 + M_1 A_2. \tag{3.67}$$

If the role of N_1 and N_2 had been permuted, the number of multiplications would be the same, but the number of additions would be $N_1^2 A_2 + M_2 A_1$. This computation mechanism can be used recursively to cover the case of more than two factors, provided all these factors are relatively prime. In practice, the convolutions of dimensions $N_1 \times N_1$ and $N_2 \times N_2$ are computed by polynomial transforms and large two-dimensional convolutions are computed from a small set of polynomial transform algorithms. A convolution of dimension 15×15, for instance, would be computed from convolutions of dimensions 3×3 and 5×5. Since these convolutions are calculated, respectively, with 13 multiplications, 70 additions and 55 multiplications, 369 additions by polynomial transforms (Table 3.4), nesting the two algorithms allows one to compute the convolution of 15×15 terms with 715 multiplications and 6547 additions. We give in Table 3.5 the number of arithmetic operations for various two-dimensional convolutions calculated by polynomial transforms and nesting. It can be seen that this approach gives a very low number of multiplications per point but that the number of additions per point is generally higher than with composite polynomial transforms.

This number of additions can be reduced by slightly refining the Agarwal–Cooley nesting method. This is done by using a split nesting technique

Table 3.6. Number of operations for two-dimensional convolutions computed by polynomial transforms and split nesting

Convolution size	Number of multiplications	Number of additions	Multiplications per point	Additions per point
12x12 (144)	286	2290	1.99	15.90
20x20 (400)	946	10826	2.36	27.06
30x30 (900)	2236	28348	2.48	31.50
36x36 (1296)	4246	34010	3.28	26.24
40x40 (1600)	4558	69534	2.85	43.46
60x60 (3600)	12298	129106	3.42	35.86
72x72 (5184)	20458	192470	3.95	37.13
80x80 (6400)	26254	308494	4.10	48.20
120x120 (14400)	59254	686398	4.11	47.67

[3.4] in which the first stage of the computation is the same as with the Agarwal–Cooley method, with a convolution of dimension $N_1 N_2 \times N_1 N_2$ calculated as a convolution of dimension $N_1 \times N_1$ in which scalars are replaced by polynomials of dimensions $N_2 \times N_2$. Thus, the A_1 additions corresponding to the algorithm $N_1 \times N_1$ yield $N_2^2 A_1$ additions as in (3.67). In the Agarwal–Cooley method, the M_1 multiplications corresponding to the algorithm $N_1 \times N_1$ are then replaced by M_1 convolutions of dimension $N_2 \times N_2$, thus yielding $M_1 A_2$ additions. In the split nesting algorithm, however, the role of N_1 and N_2 is reversed at this stage. If we assume, for instance that N_1 and N_2 are primes, the M_1 multiplications related to the algorithm $N_1 \times N_1$ correspond to $(N_1 + 1)$ polynomial products of dimension $(N_1 - 1)$ plus one multiplication. Therefore, the M_1 convolutions of dimension $N_2 \times N_2$ can be viewed as a convolution of $N_2 \times N_2$ terms in which the scalars are replaced by $(N_1 + 1)$ polynomials of $(N_1 - 1)$ terms plus one scalar. Thus, each addition in the algorithm $N_2 \times N_2$ is multiplied by N_1^2 instead of M_1 and, since $N_1^2 < M_1$, the number of additions is reduced, while the number of additions is left unchanged.

We give in Table 3.6 the number of arithmetic operations for various two-dimensional convolutions computed by polynomial transforms and split nesting. It can be seen that the number of additions is significantly lower than with conventional nesting and comparable with the approach using large composite polynomial transforms.

3.2.6 Comparison with Conventional Computation Methods

A detailed comparison of the polynomial transform approach with conventional computation methods is difficult since it depends upon the particular choice of the algorithms, the relative cost of multiplications and additions and that of ancillary operations.

For convolutions of real sequences, the polynomial transform approach requires only real arithmetic and does not use trigonometric functions. Thus, it

Table 3.7. Number of operations for two-dimensional convolutions computed by Agarwal-Cooley nesting

Convolution size	Number of multiplications	Number of additions	Multiplications per point	Additions per point
3x3 (9)	16	77	1.78	8.56
4x4 (16)	25	135	1.56	8.44
5x5 (25)	100	465	4.00	18.60
7x7 (49)	256	1610	5.22	32.86
8x8 (64)	196	1012	3.06	15.81
9x9 (81)	361	2072	4.46	25.58
12x12 (144)	400	3140	2.78	21.81
16x16 (256)	1225	7905	4.79	30.88
20x20 (400)	2500	16100	6.25	40.25
30x30 (900)	6400	41060	7.11	45.62
36x36 (1296)	9025	62735	6.96	48.41
40x40 (1600)	19600	116440	12.25	72.77
60x60 (3600)	40000	264500	11.11	73.47
72x72 (5184)	70756	488084	13.65	94.15
80x80 (6400)	122500	799800	19.14	124.97
120x120 (14400)	313600	1986240	21.78	137.93

Table 3.8. Number of real operations for real convolutions computed by FFTs. (Rader-Brenner algorithm. 2 real convolutions per DFT)

Convolution size	Number of multiplications	Number of additions	Multiplications per point	Additions per point
8x8 (64)	160	928	2.50	14.50
16x16 (256)	1024	5120	4.00	20.00
32x32 (1024)	5888	28672	5.75	28.00
64x64 (4096)	31232	147456	7.62	36.00
128x128 (16384)	156672	720896	9.56	44.00

seems logical to compare it to the Agarwal–Cooley nesting method which has also both of these characteristics and has been shown to be relatively efficient for one-dimensional convolutions. We give in Table 3.7 the number of arithmetic operations for two-dimensional convolutions computed by this method, as described in Sect. 3.2.5, but without using polynomial transforms. It can be seen by comparison with Tables 3.7 and 3.6, that the number of operations is always higher than with the polynomial transform method with the relative efficiency of this last method increasing with convolution sizes. For a convolution of 120×120, for instance, the polynomial transform approach requires about 5 times less multiplications and 2.5 times less additions than the Agarwal–Cooley method.

A comparison with the computation by FFT is somewhat more difficult since in this case, real convolutions require complex arithmetic and involve the manipulation of trigonometric functions. We give in Table 3.8 the number of real operations for real two-dimensional convolutions computed by FFT using the Rader–Brenner algorithm [3.15]. We have assumed here that two real convolutions were calculated simultaneously and that trigonometric functions were precomputed and stored. Under these conditions, which are rather favorable to the FFT approach, the number of additions is slightly larger than with the polynomial transform method while the number of multiplications is about twice that corresponding to the polynomial transform approach. Using a conventional radix-4 FFT algorithm or the Winograd Fourier transform method [3.1] would yield comparable results.

For complex convolutions, the polynomial transform approach can be implemented with two real multiplications per complex multiplications by taking advantage of the fact that $j = \sqrt{-1}$ is real in certain fields. For instance, in the case of fields of polynomials modulo $(Z^q + 1)$, with q even, $j \equiv Z^{q/2}$. Thus, a complex convolution $Y_l(Z) + j\hat{Y}_l(Z)$, defined modulo $(Z^q + 1)$ is computed with two real convolutions by

$$Q_{1,l}(Z) \equiv \sum_{m=0}^{N-1} [H_m(Z) + Z^{q/2}\hat{H}_m(Z)] [X_{l-m}(Z) + Z^{q/2}\hat{X}_{l-m}(Z)] \; \text{modulo}(Z^q + 1)$$

(3.68)

$$Q_{2,l}(Z) \equiv \sum_{m=0}^{N-1} [H_m(Z) - Z^{q/2}\hat{H}_m(Z)] [X_{l-m}(Z) - Z^{q/2}\hat{X}_{l-m}(Z)] \; \text{modulo}(Z^q + 1)$$

(3.69)

$$Y_l(Z) = [Q_{1,l}(Z) + Q_{2,l}(Z)]/2$$

(3.70)

$$\hat{Y}_l(Z) \equiv -Z^{q/2}[Q_{1,l}(Z) - Q_{2,l}(Z)]/2 \; \text{modulo} \; (Z^q + 1).$$

(3.71)

Under these conditions, complex two-dimensional convolutions are computed by polynomial transforms with about twice as many real arithmetic operations as for real convolutions and the relative efficiency of the polynomial transform approach compared with the FFT method is about the same as for real convolutions.

It should also be noted that, when the algorithms are microprogrammed, it may be advantageous to compute the short polynomial product algorithms by distributed arithmetic [3.16]. With this approach, which involves bit manipulations, the number of operations for the polynomial product algorithms is drastically reduced, thus significantly increasing the efficiency of the polynomial transform approach, at the expense of added programming complexity.

3.3 Computation of Two-Dimensional DFTs by Polynomial Transforms

We have seen that the number of arithmetic operations required for calculating two-dimensional circular convolutions is significantly reduced when conventional DFT methods are replaced by a polynomial transform approach.

Part of the savings are due to the fact that short one-dimensional convolutions and polynomial products are computed more efficiently by interpolation than by DFTs. However, the most significant factor in reducing the processing load relates to an efficient two-dimensional to one-dimensional mapping by polynomial transforms. We shall see now that these approaches can be extended to the computation of DFTs and that polynomial transforms can be used to convert two-dimensional DFTs either into one-dimensional DFTs or into one-dimensional convolutions and polynomial products [3.17]. In practice, both methods yield a significant reduction in number of arithmetic operations and can be combined in order to achieve optimum efficiency for large DFTs.

3.3.1 Reduced DFT Algorithm

We consider a two-dimensional DFT $\bar{X}_{k_1,k_2}$ of size $N \times N$

$$\bar{X}_{k_1,k_2} = \sum_{n_1=0}^{N-1} \sum_{n_2=0}^{N-1} x_{n_1,n_2} W^{n_1 k_1} W^{n_2 k_2}, \quad W = \exp -j2\pi/N \tag{3.72}$$

$$k_1, k_2 = 0, ..., N-1$$

$$j = \sqrt{-1}.$$

The conventional row-column computation method [3.18] consists in computing $\bar{X}_{k_1,k_2}$ as N one-dimensional DFTs along k_2 and N one-dimensional DFTs along k_1. In this method, $\bar{X}_{k_1,k_2}$ is mapped into $2N$ DFTs of size N and if M_1 is the number of complex multiplications required to compute this DFT, the total number M of multiplications corresponding to $\bar{X}_{k_1 k_2}$ is $M = 2NM_1$. When $M_1 < 2N$, it is more efficient to compute $\bar{X}_{k_1,k_2}$ by the Winograd nesting algorithm [3.1]. In this case, $\bar{X}_{k_1,k_2}$ is evaluated as a DFT of size N in which each scalar is replaced by a vector of N points and each multiplication is replaced by a DFT of N terms, so that $M = M_1^2$.

In order to compute $\bar{X}_{k_1 k_2}$ by polynomial transforms, we need first a representation of this DFT in polynomial algebra. This is done by replacing (3.72) by

$$\bar{X}_{k_1}(Z) \equiv \sum_{n_1=0}^{N-1} X_{n_1}(Z) W^{n_1 k_1} \quad \text{modulo } (Z^N - 1) \tag{3.73}$$

$$X_{n_1}(Z) = \sum_{n_2=0}^{N-1} x_{n_1,n_2} Z^{n_2}, \quad k_1 = 0, ..., N-1 \tag{3.74}$$

$$\bar{X}_{k_1,k_2} \equiv \bar{X}_{k_1}(Z) \quad \text{modulo } (Z - W^{k_2}), \quad k_2 = 0, ..., N-1. \tag{3.75}$$

In (3.75), $\bar{X}_{k_1,k_2}$ is obtained by substituting W^{k_2} for Z in (3.74). Thus, (3.73–75) are equivalent to (3.72). Note that $\bar{X}_{k_1}(Z)$ do not need to be defined modulo $(Z^N - 1)$ at this stage. However, this definition is valid since $Z^N \equiv W^{Nk_2} = 1$.

We assume now that N is a prime. In this case, $(Z^N - 1)$ factors into two cyclotomic polynomials with

$$Z^N - 1 = (Z - 1) P(Z) \tag{3.76}$$

$$P(Z) = Z^{N-1} + Z^{N-2} + \ldots + 1 \tag{3.77}$$

for $k_2 = 0$, which corresponds to $Z \equiv 1$, $\bar{X}_{k_1,0}$ is a simple DFT of N points

$$\bar{X}_{k_1,0} = \sum_{n_1 = 0}^{N-1} \left(\sum_{n_2 = 0}^{N-1} x_{n_1,n_2} \right) W^{n_1 k_1}, \quad k_1 = 0, \ldots, N-1 \tag{3.78}$$

for $k_2 \neq 0$, W^{k_2} is always a root of $P(Z)$

$$P(Z) = \prod_{k_2 = 1}^{N-1} (Z - W^{k_2}). \tag{3.79}$$

Since $(Z - W^{k_2})$ is a factor of $P(Z)$ and $P(Z)$ is a factor of $(Z^N - 1)$, $X_{k_1}(Z)$ can be defined modulo $P(Z)$ instead of modulo $(Z^N - 1)$ and (3.73–75) reduce to

$$\bar{X}_{1,k_1}(Z) \equiv \sum_{n_1 = 0}^{N-1} X_{1,n_1}(Z) W^{n_1 k_1} \text{ modulo } P(Z), \quad k_1 = 0, \ldots, N-1 \tag{3.80}$$

$$X_{1,n_1}(Z) = \sum_{n_2 = 0}^{N-2} (x_{n_1,n_2} - x_{n_1,N-1}) Z^{n_2} \equiv X_{n_1}(Z) \text{ modulo } P(Z) \tag{3.81}$$

$$\bar{X}_{k_1,k_2} \equiv \bar{X}_{1,k_1}(Z) \text{ modulo } (Z - W^{k_2}), \quad k_2 = 1, \ldots, N-1, \tag{3.82}$$

N being a prime and $k_2 \neq 0$, the permutation $k_1 k_2$ modulo N maps all values of k_1. Thus, indices k_1 can be permuted by multiplying by k_2, with

$$\bar{X}_{1,k_1 k_2}(Z) \equiv \sum_{n_1 = 0}^{N-1} X_{1,n_1}(Z) W^{n_1 k_1 k_2} \text{ modulo } P(Z), \quad k_1 = 0, \ldots, N-1 \tag{3.83}$$

$$\bar{X}_{k_1 k_2, k_2} \equiv \bar{X}_{1,k_1 k_2}(Z) \text{ modulo } (Z - W^{k_2}), \quad k_2 = 1, \ldots, N-1. \tag{3.84}$$

Since $\bar{X}_{k_1 k_2, k_2}$ is defined modulo $(Z - W^{k_2})$, $W^{k_2} \equiv Z$. Substituting Z for W^{k_2} in (3.83) yields

$$\bar{X}_{1,k_1 k_2}(Z) \equiv \sum_{n_1 = 0}^{N-1} X_{1,n_1}(Z) Z^{n_1 k_1} \text{ modulo } P(Z), \quad k_1 = 0, \ldots, N-1 \tag{3.85}$$

where the right-hand side of the equation is independent of k_2. $\bar{X}_{1,k_1k_2}(Z)$ is a polynomial transform of N terms which is computed without multiplications, with only $(N^3 - N^2 - 3N + 4)$ additions. Under these conditions, the only multiplications required for evaluating the DFT of $N \times N$ points are those corresponding to (3.84) and to the DFT of N points given by (3.78).

We now show that computing (3.84) is equivalent to evaluating N DFTs of N terms where the last input sample is zero and the first output sample is not computed. This can be seen by noticing that the outputs $X_{1,k_1k_2}(Z)$ of the polynomial transform (3.85) are N polynomials of $(N-1)$ terms. Assuming these polynomials are given by

$$\bar{X}_{1,k_1k_2}(Z) = \sum_{l=0}^{N-2} y_{k_1,l} Z^l \tag{3.86}$$

and substituting into (3.84) yields

$$\bar{X}_{k_1k_2,k_2} = \sum_{l=0}^{N-2} y_{k_1,l} W^{k_2 l}, \quad k_2 = 1, \ldots, N-1. \tag{3.87}$$

Thus, for N prime, the DFT of $N \times N$ points is mapped by polynomial transforms into $(N+1)$ DFTs of N terms instead of $2N$ DFTs of N terms with the row-column method. This polynomial transform approach is shown in Fig. 3.6. A slight additional reduction in number of multiplications can be obtained by taking advantage of the fact that, in the N DFTs defined by (3.87), the last input term is equal to zero and the first output term is not computed. This is done by noticing that, since $\sum_{l=1}^{N-1} W^{k_2 l} = -1$ for $k_2 \neq 0$ and N prime, (3.87) can be rewritten as

$$\bar{X}_{k_1k_2,k_2} = \sum_{l=1}^{N-1} y_{k_1,l}^1 W^{k_2 l}, \quad k_2 = 1, \ldots, N-1, \tag{3.88}$$

where

$$y_{k_1,l}^1 = (y_{k_1,l} - y_{k_1,0}), \quad y_{k_1,N-1}^1 = -y_{k_1,0}, \quad l = 1, \ldots, N-2. \tag{3.89}$$

The *reduced* DFTs (3.88) can then be computed as correlations of $(N-1)$ points by using Rader's algorithm [3.19]: If g is a primitive root modulo N [3.7], l and k_2 are redefined as

$$\begin{cases} l \equiv g^u \text{ modulo } N \\ k_2 \equiv g^v \text{ modulo } N. \end{cases} \tag{3.90}$$

Under these conditions, the reduced DFT (3.88) is converted into a correlation

$$\bar{X}_{k_1 g^v, g^v} = \sum_{u=0}^{N-2} y_{k_1,g^u}^1 W^{g^{u+v}}. \tag{3.91}$$

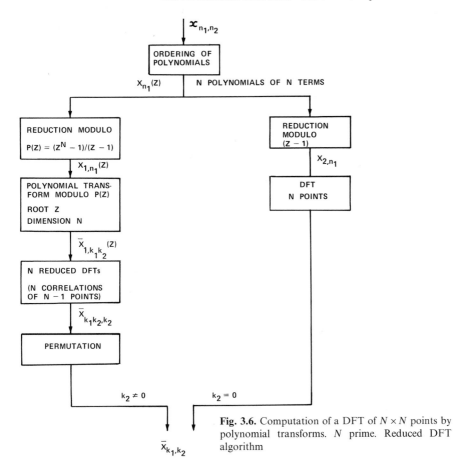

Fig. 3.6. Computation of a DFT of $N \times N$ points by polynomial transforms. N prime. Reduced DFT algorithm

The substitution of the sequence $y_{k_1,l}^1$ to $y_{k,l}$ is equivalent to a multiplication of $\bar{X}_{k_1 k_2}(Z)$ by Z^{-1} modulo $(Z^N - 1)$ followed by a reduction modulo $P(Z)$ and a multiplication by Z. In practice, the multiplication by Z^{-1} is combined with the ordering of the input polynomials. The reduction modulo $P(Z)$ is then done without additions as part of the computation of the polynomial transform.

Assuming the DFTs of N points are computed with M_1 complex multiplications by short correlation algorithms as for instance for Winograd Fourier transforms, the corresponding correlations require $(M_1 - 1)$ complex multiplications and the total number M of multiplications required to compute a DFT of $N \times N$ points by polynomial transforms reduce to

$$M = (N+1)M_1 - N. \tag{3.92}$$

This is about half that corresponding to the row-column method which would have required $2NM_1$ multiplications and always below (except for $M_1 = N$) that

Table 3.9. Computation of two-dimensional DFTs by polynomial transforms. Reduced DFT algorithm

Size NxN	Polynomial transforms	Number of additions for polynomial transforms and reductions	DFTs and reduced DFTs
N prime	1 polynomial transform of N terms modulo $(Z^N-1)/(Z-1)$	N^3+N^2-5N+4	1 DFT of N terms N reduced DFTs of N terms (N correlations of N-1 terms)
$N=q^2$ q prime	1 polynomial transform of q^2 terms modulo $(Z^{q^2}-1)/(Z-1)$ 1 polynomial transform of q terms modulo $(Z^{q^2}-1)/(Z^q-1)$	$2q^5+q^4-5q^3+q^2+6$	q^2+q reduced DFTs of q^2 terms q reduced DFTs of q terms (q correlations of q-1 terms) 1 DFT of q terms
$N=2^t$	1 polynomial transform of 2^t terms modulo $Z^{2^{t-1}}+1$ 1 polynomial transform of 2^{t-1} terms modulo $Z^{2^{t-1}}+1$	$(3t+5)2^{2(t-1)}$	3.2^{t-1} reduced DFTs of dimension N 1 DFT of dimension N/2 × N/2
$N=q_1q_2$ q_1q_2 primes	1 polynomial transform of q_1q_2 terms q_2 polynomial transforms of q_1 terms q_1 polynomial transforms of q_2 terms	$q^2_1q^2_2(q_1+q_2+2)$ $-5q_1q_2(q_1+q_2)$ $+4(q^2_1+q^2_2)$	$q_1q_2+q_1+q_2$ reduced DFTs of q_1q_2 terms q_1 reduced DFTs of q_1 terms q_2 reduced DFTs of q_2 terms 1 DFT of q_1q_2 terms

corresponding to the Winograd nesting algorithm which is M_1^2. Moreover, when the DFTs and reduced DFTs of N terms are computed by short correlation algorithms, all complex multiplications reduce to multiplications by pure real or pure imaginary numbers and can be implemented with only two real multiplications.

Similar polynomial transform approaches to DFT computation can also easily be defined for N composite. The existence conditions are established by a method very similar to that used in Sect. 3.2.1 for two-dimensional con-

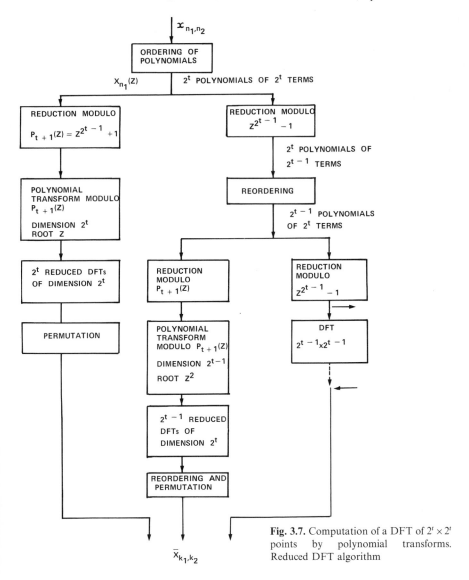

Fig. 3.7. Computation of a DFT of $2^t \times 2^t$ points by polynomial transforms. Reduced DFT algorithm

volutions and are summarized in Table 3.9. A particularly interesting case corresponds to DFTs of $N \times N$ points, with $N = 2^t$. These DFTs are computed as shown in Fig. 3.7 with one polynomial transform of 2^t terms modulo $P_{t+1}(Z) = Z^{2^{t-1}} + 1$, one polynomial transform of 2^{t-1} terms modulo $P_{t+1}(Z)$, $2^t + 2^{t-1}$ reduced DFTs of dimension 2^t with half their input and output terms being null, and one DFT of $N/2 \times N/2$ points. The DFT of $N/2 \times N/2$ points can in turn be computed by a similar process. The main interest of the approach corresponding to $N = 2^t$ stems from the fact that the polynomial transforms are computed with an FFT-type algorithm so that all calculations are organized in a way very similar to conventional FFT computation.

Moreover, the reduced DFTs of dimension N can then be viewed as conventional DFTs of dimension $N/2$, where the input sequence is multiplied pointwise by the sequence 1, W, W^2 and are therefore identical to the reduced DFTs that appear in the first stage of a decimation in frequency, radix-2 FFT decomposition. More precisely, the reduced DFT of N points, for $N = 2^t$ is given by

$$\bar{X}_{(2v+1)k_1, 2v+1} = \sum_{l=0}^{N/2-1} y_{k_1, l} W^l W^{2vl} \qquad v = 0, \ldots, N/2 - 1. \tag{3.93}$$

In this reduced DFT, the multiplications by W^l are normally complex. However, these multiplications can be reduced to multiplications by pure real or pure imaginary numbers by using the Rader–Brenner algorithm [3.15] or one of its derivatives. This is done by substituting to the input sequence $y_{k_1, l}$, the sequence $y_{k_1, l}^1$ defined by

$$\begin{cases} y_{k_1, 0}^1 = 0, \quad y_{k_1, N/4}^1 = 0 \\ y_{k_1, l}^1 = y_{k_1, l}/2 \cos(2\pi l/N), \quad l \neq 0, N/4. \end{cases} \tag{3.94}$$

The $N/2$ points DFT $\bar{X}_{(2v+1)k_1, 2v+1}^1$ of $y_{k_1, l}^1$ is then computed with

$$\bar{X}_{(2v+1)k_1, 2v+1}^1 = \sum_{l=0}^{N/2-1} y_{k_1, l}^1 W^{2vl}. \tag{3.95}$$

Since

$$W^{2vl} + W^{2(v+1)l} = W^l W^{2vl}(W^l + W^{-l}) = 2W^l W^{2vl} \cos(2\pi l/N),$$

$\bar{X}_{(2v+1)k_1, 2v+1}$ is obtained by simple additions

$$\begin{cases} \bar{X}_{(2v+1)k_1, 2v+1} = \bar{X}_{(2v+1)k_1, 2v+1}^1 + \bar{X}_{(2v+3)k_1, 2v+3}^1 + e_0, & v \text{ even} \\ \bar{X}_{(2v+1)k_1, 2v+1} = \bar{X}_{(2v+1)k_1, 2v+1}^1 + \bar{X}_{(2v+3)k_1, 2v+3}^1 + e_1, & v \text{ odd} \end{cases} \tag{3.96}$$

where

$$e_0 = y_{k_1, 0} - j y_{k_1, N/4}, \qquad e_1 = y_{k_1, 0} + j y_{k_1, N/4}.$$

Using this method, the reduced DFT defined by (3.93) is calculated with $(M_1 + N - 4)$ real multiplications and $A_1 + 2N$ real additions, where M_1 and A_1 are, respectively, the number of multiplications and additions required to evaluate a DFT of $N/2$ points.

We give in Sect. 3.6 various short reduced DFT algorithms and we summarize in Table 3.10 the number of real operations for the most frequently used algorithms. Using these data, short DFT algorithms designed for the

Table 3.10. Number of real operations for short DFTs and reduced DFTs. Trivial multiplications by ± 1, $\pm j$ are given between parentheses

Size N	Number of multiplications	Number of additions	
2	4 (4)	4	
3	6 (2)	12	
4	8 (8)	16	
5	12 (2)	34	
7	18 (2)	72	
8	16 (12)	52	
9	22 (2)	88	DFTs
16	36 (16)	148	
32	104 (36)	424	
64	272 (76)	1104	
128	672 (156)	2720	
256	1600 (316)	6464	
512	3712 (636)	14976	
1024	8448 (1276)	34048	
3	4 (0)	8	Reduced DFTs
5	10 (0)	30	computed as
7	16 (0)	68	correlations
4	4 (4)	4	
8	8 (4)	20	
9	16 (0)	56	
16	20 (4)	64	
32	68 (20)	212	
64	168 (40)	552	Reduced DFTs
128	400 (80)	1360	
256	928 (160)	3232	
512	2112 (320)	7488	
1024	4736 (640)	17024	

Table 3.11. Number of real operations for complex two-dimensional DFTs computed by polynomial transforms with the reduced DFT algorithm. Trivial multiplications by ± 1, $\pm j$ are not counted

DFT size	Number of multiplications	Number of additions	Multiplications per point	Additions per point
2x2	0	16	0	4.00
3x3	16	86	1.78	9.56
4x4	0	128	0	8.00
5x5	60	442	2.40	17.68
7x7	128	1270	2.61	25.92
8x8	48	816	0.75	12.75
9x9	208	1570	2.57	19.38
16x16	432	4528	1.69	17.69
32x32	2736	24944	2.67	24.36
64x64	15024	125040	3.67	30.53
128x128	76464	599152	4.67	36.57
256x256	371376	2790512	5.67	42.58
512x512	1747632	12735600	6.67	48.58
1024x1024	8039088	57234544	7.67	54.58

Winograd Fourier transform [3.1, 21] and Rader–Brenner algorithms, we give in Table 3.11 the number of real operations for various two-dimensional DFTs.

A detailed comparison with conventional computation methods will be given in Sect. 3.3.3, but we can note at this stage that the number of multiplications is about half that corresponding to conventional FFT computation using the Rader–Brenner algorithm while the number of additions is somewhat reduced. The number of operations is also smaller than with conventional Winograd Fourier transforms.

3.3.2 Nesting and Prime Factor Algorithms

The polynomial transform approach discussed in the preceding section allows one to compute large two-dimensional DFTs of dimension $N \times N$ provided large reduced DFT algorithms are available. For $N = 2^t$, these reduced DFT algorithms are easily computed by the Rader–Brenner method. When N is composite and not a power of two, large reduced DFT algorithms can be computed by nesting small reduced DFT algorithms. However, when N is composite, it is often more convenient to construct large transforms by using a small set of short transforms in a nesting [3.1] or prime factor algorithm [3.20, 21]. These approaches are particularly well adapted when the DFT has mutually prime factors in both dimensions. In the following, we will restrict our discussion to DFTs of size $N_1 N_2 \times N_1 N_2$, where N_1 and N_2 are relatively prime, since larger DFTs, with more than two relatively prime factors, can easily be computed recursively from the two-factors algorithm.

In the nesting algorithm, the two-dimensional DFT of dimension $N_1 N_2 \times N_1 N_2$ is first converted into a four-dimensional DFT of size $(N_1 \times N_1) \times (N_2 \times N_2)$ by a simple index mapping, based on the Chinese remainder theorem and similar to that described in Sect. 3.2.5 for two-dimensional convolutions. The four-dimensional DFT is in turn computed with Winograd nesting [3.1] by calculating by polynomial transforms a DFT of dimension $N_1 \times N_1$ in which each scalar is replaced by an array of $N_2 \times N_2$ terms and each multiplication is replaced by a DFT of $N_2 \times N_2$ points computed by polynomial transforms. Thus, if M_1, M_2, M and A_1, A_2, A are, respectively, the number of complex multiplications and additions required to evaluate DFTs of sizes $N_1 \times N_1$, $N_2 \times N_2$ and $N_1 N_2 \times N_1 N_2$, we have

$$M = M_1 M_2 \tag{3.97}$$

$$A = N_2^2 A_1 + M_1 A_2. \tag{3.98}$$

The four-dimensional DFT of dimension $(N_1 \times N_1) \times (N_2 \times N_2)$ can also be computed by a prime factor algorithm or row-column method as N_1^2 DFTs of dimension $N_2 \times N_2$ and N_2^2 DFTs of dimension $N_1 \times N_1$. In this case, the number of operations becomes

$$M = N_1^2 M_2 + N_2^2 M_1 \tag{3.99}$$

$$A = N_1^2 A_2 + N_2^2 A_1 . \tag{3.100}$$

Since $M_1 \geqq N_1^2$, $M_2 \geqq N_2^2$, the nesting method always requires more additions than the prime factor algorithm. However, if M_1 and M_2 are not much larger than N_1^2 and N_2^2, which is the case for relatively short DFTs, the nesting method requires less multiplications than the prime factor algorithms for a number of additions which is only slightly increased. Thus, the nesting algorithm is generally better suited for DFTs of moderate sizes while the prime factor technique is best for large DFTs.

The computing efficiency of these algorithms can be improved by splitting the calculations [3.17]. In the case of the nesting method, this approach reduces the number of additions in a way similar to that described in Sect. 3.2.5 for convolutions. For the prime factor algorithm, splitting the calculations reduces the number of multiplications. In order to simplify the notations, we will describe here this last algorithm in the case of a DFT of $N_1 \times N_2$ points, the results being easily extended to a DFT of $(N_1 \times N_1) \times (N_2 \times N_2)$ points by replacing the scalars by vectors of N_1 and N_2 terms, respectively. In the following, we will assume that N_1 and N_2 are primes. The DFT $\bar{X}_{k_1,k_2}$ of $N_1 \times N_2$ terms is given by

$$\bar{X}_{k_1,k_2} = \sum_{n_1=0}^{N_1-1} W_1^{n_1 k_1} \sum_{n_2=0}^{N_2-1} x_{n_1,n_2} W_2^{n_2 k_2} \tag{3.101}$$

$$W_1 = \exp{-j2\pi/N_1}, \quad W_2 = \exp{-j2\pi/N_2}, \quad \begin{array}{l} k_1 = 0, ..., N_1 - 1 \\ k_2 = 0, ..., N_2 - 1 . \end{array}$$

By using Rader's algorithm [3.19], this DFT is computed as one DFT of N_1 points, one correlation of $(N_1 - 1) \times (N_2 - 1)$ points and one correlation of $(N_2 - 1)$ points

$$\bar{X}_{k_1,0} = \sum_{n_1=0}^{N_1-1} \left(\sum_{n_2=0}^{N_2-1} x_{n_1,n_2} \right) W_1^{n_1 k_1}, \quad k_1 = 0, ..., N_1 - 1 \tag{3.102}$$

$$\bar{X}_{0,g^{v_2}} = \sum_{u_2=0}^{N_2-2} \left[\sum_{n_1=0}^{N_1-1} (x_{n_1,g^{u_2}} - x_{n_1,0}) \right] W_2^{g^{u_2+v_2}} \tag{3.103}$$

$$\bar{X}_{h^{v_1},g^{v_2}} = \sum_{u_2=0}^{N_2-2} W_2^{g^{u_2+v_2}} \sum_{u_1=0}^{N_1-2} (x_{h^{u_1},g^{u_2}} - x_{h^{u_1},0} - x_{0,g^{u_2}} + x_{0,0}) W_1^{h^{u_1+v_1}} \tag{3.104}$$

where h and g are primitive roots modulo N_1 and N_2 with

$$k_1 \equiv h^{v_1} \text{ modulo } N_1, n_1 \equiv h^{u_1} \text{ modulo } N_1, \quad u_1, v_1 = 0, ..., N_1 - 2 \tag{3.105}$$

$$k_2 \equiv g^{v_2} \text{ modulo } N_2, n_2 \equiv g^{u_2} \text{ modulo } N_2, \quad u_2, v_2 = 0, ..., N_2 - 2. \tag{3.106}$$

The two-dimensional correlation (3.104) is half separable, and can therefore be computed as $(N_2 - 1)$ correlations of $(N_1 - 1)$ points plus $(N_1 - 1)$ correlations of $(N_2 - 1)$ points. Under these conditions, if M_1 and M_2 are the number of multiplications corresponding to the DFTs of N_1 and N_2 points, the total number of multiplications reduces to $N_1 M_2 + N_2 M_1 - N_1 - N_2 + 1$ instead of $N_1 M_2 + N_2 M_1$ for the conventional prime factor algorithm. Thus, splitting the computation saves $N_1 + N_2 - 1$ multiplications. Larger savings can be achieved by reducing the correlations into cyclotomic polynomials.

3.3.3 Computation of Winograd Fourier Transforms by Polynomial Transforms

In Sect. 3.3.1, we have discussed an efficient method of computing DFTs by polynomial transforms. This reduced DFT technique is applicable mainly to multidimensional DFTs having a common factor in two or more dimensions and is based on a mapping of a multidimensional DFT into one-dimensional DFTs and reduced DFTs by means of polynomial transforms. In this section, we present another way of computing DFTs by polynomial transforms [3.1, 17]. We will show that, when a DFT is computed by the Winograd algorithm, polynomial transforms can be used to map DFTs into one-dimensional convolutions and polynomial products and allow significant processing savings over the Winograd algorithm used alone.

In this method, the DFT is first converted into multidimensional correlations by using the Winograd algorithm [3.1, 21]. In the case of a DFT of dimension $N_1 \times N_2$, this is done by computing a DFT of dimension N_1 in which each scalar is replaced by a DFT of N_2 terms. If the DFTs of dimensions N_1 and N_2 are calculated as correlations by Rader's algorithm, this computation process can be viewed as converting the DFT of dimension $N_1 \times N_2$ into one-dimensional correlations and DFTs and two-dimensional correlations. In the simple case corresponding to N_1 and N_2 primes, for instance, the DFT of $N_1 \times N_2$ points is computed as one DFT of N_1 points, one correlation of $(N_2 - 1)$ points and one correlation of dimension $(N_1 - 1) \times (N_2 - 1)$ as shown by (3.102–104). The same method is used to evaluate a one-dimensional DFT of dimension $N_1 N_2$ when N_1 and N_2 are relatively prime, since this DFT can be converted into a two-dimensional DFT of dimension $N_1 \times N_2$ by a simple index mapping using Good's algorithm [3.20]. The same technique can also be applied recursively to cover the case of more than two factors. In this case, the DFT is converted into multidimensional correlations.

In the conventional Winograd algorithm, the multidimensional correlations are calculated by a simple nesting of the short one-dimensional correlations, which is equivalent to computing these correlations by the Agarwal–Cooley algorithm [3.2]. We have seen however in Sect. 3.2 that multidimensional correlations are computed more efficiently by polynomial transforms than by the Agarwal–Cooley method when these correlations have common factors in

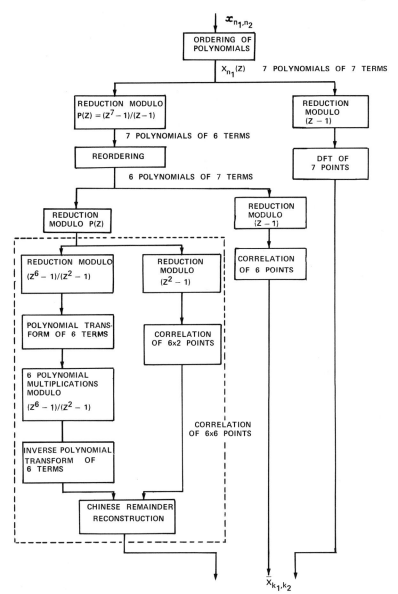

Fig. 3.8. Computation of a DFT of 7×7 points by the Winograd algorithm and polynomial transforms

several dimensions. Thus, computing multidimensional correlations by polynomial transforms reduces the number of additions and multiplications in the Winograd Fourier transform algorithm.

This method is illustrated in Fig. 3.8 for a DFT of dimension 7×7. Since the DFT of 7 terms is computed with one multiplication and one correlation of 6 terms, the DFT of 7×7 points is mapped into one DFT of 7 terms, one correlation of 6 terms and one correlation of 6×6 terms. Therefore, if we use an algorithm requiring 8 complex multiplications for the correlation of 6 terms, the DFT of dimension 7×7 is calculated with 81 complex multiplications with the conventional Winograd algorithm. Computing the correlation of 6×6 terms by polynomial transforms requires only 52 multiplications instead of 64 with conventional nesting. The number of complex multiplications required to compute the DFT of dimension 7×7 is therefore reduced to 69 instead of 81 when the polynomial transform approach is combined with the Winograd algorithm. It should be noted however, that the first polynomial transform method, which is based on reduced DFTs would have required in this case only 65 complex multiplications. In fact, this result is quite general and the first polynomial transform method, which operates over larger fields is usually preferable whenever it is applicable.

The main interest of the second polynomial transform method, which is based on smaller extension fields, lies in the calculation of DFTs where the first method is not applicable. This is the case for two-dimensional DFTs when both dimensions have no common factors while the corresponding two-dimensional correlation has common factors in both dimensions. A DFT of 7×9, for instance, cannot be computed efficiently by the first method, since 7 and 9 have no common factors. By using the Winograd algorithm, this DFT is computed by nesting a DFT of 7 terms with a DFT of 9 terms. Using Rader's algorithm, the DFT of 7 terms is converted into one multiplication and one correlation of 6 terms while the DFT of 9 terms is converted into 5 multiplications and one correlation of 6 terms. Thus, the DFT of dimension 7×9 is computed with five DFTs of 7 terms, one correlation of 6 terms and one correlation of 6×6 terms. Using polynomial transforms to compute the correlation of dimension 6×6 yields an algorithm with only 174 real multiplications as opposed to 198 real multiplications for the conventional Winograd algorithm.

For large, multidimensional DFTs, it is advantageous to combine the two polynomial transform methods in order to achieve maximum computing efficiency. If we apply this technique to a DFT of 63×63 points, for instance, this DFT is calculated by nesting DFTs of dimensions 7×7 and 9×9 evaluated by the first polynomial transform method which maps the DFT of 7×7 points into 1 multiplication, plus 8 correlations of 6 terms and the DFT of 9×9 points into 33 multiplications plus 12 correlations of 6 terms. Thus, the DFT of dimension 63×63 is computed with 33 multiplications, 276 correlations of 6 terms and 96 correlations of dimension 6×6. If the correlations of dimension 6×6 are computed by polynomial transforms, the DFT of dimension 63×63 is calculated with only 11,344 real multiplications as opposed to 13,650 multiplications when the first method is used alone and 19,602 multiplication for the conventional Winograd algorithm. It should be noted that, when the two polynomial transform methods are combined, the number of additions is also lower than when the first method is used alone.

We give in Table 3.12 the number of real operations for complex two-dimensional DFTs computed by combining the two polynomial transform

Table 3.12. Number of real operations for complex two-dimensional DFTs computed by combining the two polynomial transform methods and using the split nesting technique

DFT size	Number of multiplications	Number of additions	Multiplications per point	Additions per point
63x63	11344	193480	2.86	48.75
80x80	16944	231344	2.65	36.15
120x120	35632	553392	2.47	38.43
240x240	153904	2542896	2.67	44.15
504x504	726064	15621424	2.86	61.50
1008x1008	3449024	71455456	3.39	70.33

Table 3.13. Number of real operations for complex two-dimensional DFTs computed by the Rader-Brenner FFT algorithm and the row-column method

DFT size	Number of multiplications	Number of additions	Multiplications per point	Additions per point
8x8	64	832	1.00	13.00
16x16	640	4736	2.50	18.50
32x32	4352	27136	4.25	26.50
64x64	25088	141312	6.12	34.50
128x128	132096	696320	8.06	42.50
256x256	657408	3309568	10.03	50.50
512x512	3149824	15335424	12.02	58.50
1024x1024	14688256	69730304	14.01	66.50

methods. It can be seen that, even for large transforms, the number of multiplications and additions can be very small, as reflected by the fact that a DFT of dimension 1008×1008 is calculated with 3.39 real multiplications per point or about one complex multiplication per point.

In practice, the computational savings obtained with the polynomial transform methods can be very significant as can be seen by comparing the figures given in Tables 3.11 and 3.12 with those given in Table 3.13 and corresponding to two-dimensional DFTs computed by the Rader–Brenner FFT algorithm and the row-column method. When the first polynomial transform method is used alone, the number of additions is always smaller than with the FFT algorithm and the number of multiplications is reduced by a factor of about 2 for large DFTs. When the two polynomial transform methods are combined the number of additions is about the same as for the FFT algorithm and the number of multiplications is reduced by a factor of about 4 for large DFTs. The polynomial transform approach is also significantly more efficient than the Winograd–Fourier transform algorithm used alone, as can be seen by noting that a DFT of 1008×1008 points computed by this method requires 6.25 multiplications and 91.61 additions per point. This contrasts with the first polynomial transform method, which requires 7.67 multiplications and 54.58 additions per point for a DFT of dimension 1024×1024 and the combination of the two polynomial transform approaches which yields an algorithm with only 3.39 multiplications and 70.33 additions per point for a DFT of dimension 1008×1008.

3.3.4 Relationship Between Polynomial Transforms and DFTs

An examination of (3.28–31) reveals that the existence of a polynomial transform of dimension N having the circular convolution property depends only on the existence of root of unity of order N and the existence of N^{-1}. These existence conditions are common to all transforms of dimension N having the circular convolution property [3.22] and an N term-polynomial transform having a root Z and defined modulo $P(Z)$ can be viewed as a DFT defined in the ring of polynomials modulo $P(Z)$.

This is similar to number theoretic transforms which can be considered as DFTs defined in a ring of integers. In fact, NTTs are particular cases of polynomial transforms in which the N bits words are to be viewed as polynomials. This is particularly apparent in the case of a polynomial transform of dimension 2^{t+1} and defined modulo $(Z^{2^t}+1)$. Such a transform computes a circular convolution of dimension 2^{t+1} on polynomials of dimension 2^t. If the 2^{t+1} input polynomials are defined as words of 2^t bits, the polynomial transform reduces to a Fermat number transform of dimension 2^{t+1}, of root 2 and defined modulo $(2^{2^t}+1)$.

Since polynomial transforms are DFTs defined in rings of polynomials, they are essentially equivalent to multidimensional DFTs so that their primary application is in the solution of multidimensional problems. Their main interest over DFTs and NTTs stems from the fact that systematic advantage is taken from the operation in a ring of polynomials to define simple roots of unity which allows one to eliminate the multiplications for transform computation. For some applications, like the computation of convolutions of dimension $N \times N$, polynomial transforms are optimum in the sense that they can yield algorithms with a number of general multiplications that cannot be further reduced by purely algebraic means.

3.4 Concluding Remarks

Polynomial transforms provide an efficient means of mapping two-dimensional convolutions and DFTs into one-dimensional convolutions and DFTs. These transforms are particularly interesting when the two-dimensional convolutions and DFTs have a common factor in both dimensions. In these cases, polynomial transforms yield significant savings in number of arithmetic operations over more conventional computation methods.

When computed by polynomial transforms, two-dimensional convolutions and DFTs reduce to one-dimensional convolutions and DFTs. These one-dimensional convolutions and DFTs can in turn be calculated by a number of different methods such as DFTs, NTTs, distributed arithmetic or short convolution algorithms, giving a large variety of different algorithms in which it is always possible to balance the number of multiplications and additions in order to fit a particular application.

Perhaps the most interesting polynomial transforms are those having a dimension which is a power of two. These transforms have a structure which is similar to the FFT and can therefore be programmed in a similar fashion. By combining these transforms with conventional FFT algorithms for the computation of one-dimensional convolutions, polynomial products and DFTs, one obtains two-dimensional algorithms which retain the basic simplicity of the FTT approach while allowing a significant reduction in number of operations.

3.5 Appendix – Short Polynomial Product Algorithms

Short polynomial product algorithms are given here. These algorithms have been designed to minimize the number of multiplications while keeping reasonably low the number of additions. The input sequences are labelled $\{x_n\}$ and $\{h_l\}$. $\{h_l\}$ is supposed to be constant, with expressions involving $\{h_l\}$ being precomputed and stored. The output sequence is labelled $\{y_m\}$. Expressions written between parentheses indicate groupings of additions. Input and output additions must be executed in the specified index numerical order.

3.5.1 Polynomial Product Modulo (Z^2+1)

3 multiplications, 3 additions

$$a_0 = x_0 + x_1 \qquad\qquad b_0 = h_0$$
$$a_1 = x_1 \qquad\qquad b_1 = h_0 + h_1$$
$$a_2 = x_0 \qquad\qquad b_2 = h_1 - h_0$$

$$m_k = a_k b_k \qquad k = 0, ..., 2$$

$$y_0 = m_0 - m_1$$
$$y_1 = m_0 + m_2$$

3.5.2 Polynomial Product Modulo $(Z^3-1)/(Z-1)$

3 multiplications, 3 additions

$$a_0 = x_1 \qquad\qquad b_0 = h_0 - h_1$$
$$a_1 = x_0 - x_1 \qquad\qquad b_1 = h_0$$
$$a_2 = x_0 \qquad\qquad b_2 = h_1$$

$$m_k = a_k b_k \qquad k = 0, 1, 2$$

$$y_0 = m_0 + m_1$$
$$y_1 = m_0 + m_2$$

3.5.3 Polynomial Product Modulo $(Z^4 + 1)$

9 multiplications, 15 additions

$$a_0 = x_1 + x_3 \qquad\qquad b_0 = h_0 - h_3$$
$$a_1 = (x_0 + x_2) - (x_1 + x_3) \qquad b_1 = h_0$$
$$a_2 = (x_0 + x_2) \qquad\qquad b_2 = h_0 + h_1$$
$$a_3 = x_3 \qquad\qquad\qquad b_3 = h_0 + h_2 + h_1 - h_3$$
$$a_4 = x_2 - x_3 \qquad\qquad b_4 = h_0 + h_2$$
$$a_5 = x_2 \qquad\qquad\qquad b_5 = h_0 + h_2 + h_1 + h_3$$
$$a_6 = x_1 \qquad\qquad\qquad b_6 = -h_0 + h_2 + h_1 + h_3$$
$$a_7 = x_0 - x_1 \qquad\qquad b_7 = -h_0 + h_2$$
$$a_8 = x_0 \qquad\qquad\qquad b_8 = -h_0 + h_2 - h_1 + h_3$$

$$m_k = a_k b_k \qquad k = 0, \ldots, 8$$

$$y_0 = (m_0 + m_1) - (m_3 + m_4)$$
$$y_1 = (m_2 - m_1) + (m_4 - m_5)$$
$$y_2 = (m_0 + m_1) + (m_6 + m_7)$$
$$y_3 = (m_2 - m_1) + (m_8 - m_7)$$

3.5.4 Polynomial Product Modulo $(Z^5 - 1)/(Z - 1)$

9 multiplications, 16 additions

$$a_0 = x_0 \qquad\qquad\qquad b_0 = h_0$$
$$a_1 = x_1 \qquad\qquad\qquad b_1 = h_1$$
$$a_2 = x_0 - x_1 \qquad\qquad b_2 = -h_0 + h_1$$
$$a_3 = x_2 \qquad\qquad\qquad b_3 = h_2$$
$$a_4 = x_3 \qquad\qquad\qquad b_4 = h_3$$
$$a_5 = x_2 - x_3 \qquad\qquad b_5 = -h_2 + h_3$$
$$a_6 = x_0 - x_2 \qquad\qquad b_6 = h_2 - h_0$$
$$a_7 = x_1 - x_3 \qquad\qquad b_7 = h_3 - h_1$$
$$a_8 = -a_6 + a_7 \qquad\qquad b_8 = b_6 - b_7$$

$$m_k = a_k b_k \qquad k = 0, \ldots, 8$$

$$u_0 = m_0 - m_7 \qquad\qquad u_1 = m_2 + m_0$$

$$y_0 = u_0 - m_1 + m_5$$
$$y_1 = u_1 - m_3 - m_7$$
$$y_2 = u_0 - m_4 + m_6$$
$$y_3 = u_1 + m_5 + m_6 + m_8$$

3.5.5 Polynomial Product Modulo $(Z^9-1)/(Z^3-1)$

15 multiplications, 39 additions

$a_0 = x_0 + x_2$
$a_1 = x_3 + x_5$
$a_2 = a_1 + x_4$
$a_3 = a_0 + x_1$
$a_4 = a_3 - a_2$
$a_5 = a_1 - x_4$
$a_6 = a_0 - x_1$
$a_7 = a_6 - a_5$
$a_8 = x_3$
$a_9 = x_0 - x_3$
$a_{10} = x_0$
$a_{11} = x_5$
$a_{12} = x_2 - x_5$
$a_{13} = x_2$
$a_{14} = -a_{12} + x_0 - x_4$
$a_{15} = a_9 + x_5 - x_1$
$a_{16} = -a_{15} + a_{14}$

$b_2 = (h_0 + 3h_1 + 2h_2 - 2h_3 - 3h_4 - h_5)/6$
$b_3 = (h_0 - h_2 + h_3 + 3h_4 + 2h_5)/6$
$b_4 = b_2 + b_3$
$b_5 = (-h_0 + h_1 - h_4 + h_5)/2$
$b_6 = (h_0 - h_2 - h_3 + h_4)/2$
$b_7 = b_5 + b_6$
$b_8 = 2h_0 + h_1 - h_2 - 2h_3 + h_5$
$b_9 = 2h_0 - h_2 + h_4$
$b_{10} = b_9 - b_8$
$b_{11} = h_0 - h_1 - 2h_2 + h_4$
$b_{12} = -h_1 + h_3 - 2h_5$
$b_{13} = b_{12} - b_{11}$
$b_{14} = (h_0 - h_2 - 2h_3 + 2h_5)/3$
$b_{15} = (-h_0 + h_2 - h_3 + h_5)/3$
$b_{16} = b_{15} - b_{14}$

$$m_k = a_{k+2}b_{k+2} \qquad k = 0, \ldots, 14$$

$u_0 = m_0 + m_1$
$u_1 = m_3 + m_4$
$u_2 = m_{13} + m_{14}$
$u_3 = u_0 + u_1$
$u_4 = m_0 + m_2$
$u_5 = m_3 + m_5$
$u_6 = m_{12} + m_{14}$

$u_7 = -u_3 + m_6$
$u_8 = u_4 + u_5$
$u_9 = m_9 - u_6$

$y_0 = m_7 + u_2 + u_7$
$y_1 = u_8 + m_{10} + u_9$
$y_2 = u_4 - u_5 + u_2$
$y_3 = u_7 + u_8 + m_8 + u_6$
$y_4 = u_3 + m_{11} + u_9 + u_2$
$y_5 = u_0 - u_1 + u_6$

3.5.6 Polynomial Product Modulo $(Z^7 - 1)/(Z - 1)$

15 multiplications, 53 additions

$a_0 = x_0 + x_2$
$a_1 = a_0 + x_1$
$a_2 = a_1 + x_2$
$a_3 = x_3 + x_5$
$a_4 = a_3 + x_4$
$a_5 = a_4 + x_5$
$a_6 = x_0$
$a_7 = a_1$
$a_8 = a_0 - x_1$
$a_9 = a_2 + a_2 - x_0$
$a_{10} = x_2$
$a_{11} = x_3$
$a_{12} = a_4$
$a_{13} = a_3 - x_4$
$a_{14} = a_5 + a_5 - x_3$
$a_{15} = x_5$
$a_{16} = a_{11} - a_6$
$a_{17} = a_{12} - a_7$
$a_{18} = a_{13} - a_8$
$a_{19} = a_{14} - a_9$
$a_{20} = a_{15} - a_{10}$

$b_0 = (-2h_5 + 3h_4 - h_3 - 2h_2 + h_1 + 2h_0)/2$
$b_1 = (3h_5 - 11h_4 + 10h_3 + 3h_2 - 11h_1 - 4h_0)/14$
$b_2 = (3h_5 - h_4 - 2h_3 + 3h_2 - h_1)/6$
$b_3 = (h_4 - h_3 + h_1)/6$
$b_4 = -h_5 - 2h_4 + 3h_3 - h_2 - 2h_1 + h_0$
$b_5 = (h_5 + 2h_4 - h_3 - 2h_2 + 3h_1 - h_0)/2$
$b_6 = (-11h_5 - 4h_4 + 10h_3 + 3h_2 - 11h_1 + 10h_0)/14$
$b_7 = (-h_5 - 2h_3 + 3h_2 - h_1 - 2h_0)/6$
$b_8 = (h_5 - h_3 + h_1 - h_0)/6$
$b_9 = -2h_5 + h_4 + 2h_3 - h_2 - 2h_1 + 3h_0$
$b_{10} = (2h_4 - h_3 - 2h_2 + h_1)/2$
$b_{11} = (-2h_5 - 2h_4 + 12h_3 + 5h_2 - 9h_1 - 2h_0)/14$
$b_{12} = (-2h_3 + 3h_2 - h_1)/6$
$b_{13} = (-h_3 + h_1)/6$
$b_{14} = 2h_3 - h_2 - 2h_1 + h_0$

$$m_k = a_{k+6}b_k \qquad k = 0, \ldots, 14$$

$u_0 = m_5 + m_0$
$u_1 = m_6 + m_1$
$u_2 = m_7 + m_2$
$u_3 = m_8 + m_3$
$u_4 = m_9 + m_4$
$u_5 = m_{10} + m_0$
$u_6 = m_{11} + m_1$
$u_7 = m_{12} + m_2$
$u_8 = m_{13} + m_3$
$u_9 = m_{14} + m_4$
$u_{10} = u_1 + u_3$
$u_{11} = u_{10} + u_2$
$u_{12} = u_0 + u_{11}$

$u_{13} = u_{10} + u_3$
$u_{14} = u_{13} - u_2$
$u_{15} = u_{13} + u_3 + u_3 + u_4 + u_2$
$u_{16} = -u_{12} - u_{15} - u_{14}$
$u_{17} = u_7 - u_8$
$u_{18} = u_5 + u_{17} + u_7$
$u_{19} = u_{17} - u_6 - u_8$
$u_{20} = (u_7 + u_8) + (u_7 + u_8) + u_9$
$u_{21} = u_{19} + u_{19}$
$u_{22} = u_{21} - u_{18}$
$u_{23} = u_{21} - u_{20}$

$y_0 = u_{18}$
$y_1 = u_{16} + u_{19}$
$y_2 = u_{15} + u_{23}$

$y_3 = u_{14} + u_{21}$
$y_4 = u_{12} + u_{22}$
$y_5 = u_{20}$

3.5.7 Polynomial Product Modulo ($Z^8 + 1$)

21 multiplications, 77 additions

$a_0 = x_0 + x_2$
$a_1 = x_1 + x_3$
$a_2 = x_0 - x_2$
$a_3 = x_7 - x_5$
$a_4' = x_4 + x_6$
$a_5 = x_5 + x_7$
$a_6 = x_4 - x_6$
$a_7 = x_1 - x_3$
$a_8 = a_0 + a_1$ $b_0 = (h_0 + h_1 + h_2 - h_3 + h_4 + h_5 + h_6 + h_7)/4$
$a_9 = a_4 + a_5$ $b_1 = (-h_0 - h_1 - h_2 - h_3 + h_4 + h_5 + h_6 - h_7)/4$
$a_{10} = a_8 + a_9$ $b_2 = (h_3 - h_4 - h_5 - h_6)/4$
$a_{11} = a_0 - a_1$ $b_3 = (5h_0 - 5h_1 + 5h_2 - 7h_3 + 5h_4 - 5h_5 + 5h_6 - h_7)/20$
$a_{12} = a_4 - a_5$ $b_4 = (-5h_0 + 5h_1 - 5h_2 + h_3 + 5h_4 - 5h_5 + 5h_6 - 7h_7)/20$
$a_{13} = a_{11} + a_{12}$ $b_5 = (3h_3 - 5h_4 + 5h_5 - 5h_6 + 4h_7)/20$
$a_{14} = a_2 + a_3$ $b_6 = (h_0 + h_1 - h_2 - h_3 + h_4 - h_5 - h_6 + 3h_7)/4$
$a_{15} = a_6 + a_7$ $b_7 = (-h_0 + h_1 + h_2 - 3h_3 + h_4 + h_5 - h_6 - h_7)/4$
$a_{16} = a_{14} + a_{15}$ $b_8 = (-h_1 + 2h_3 - h_4 + h_6 - h_7)/4$
$a_{17} = a_2 - a_3$ $b_9 = (5h_0 - 5h_1 - 5h_2 + h_3 + 5h_4 + 5h_5 - 5h_6 - 3h_7)/20$
$a_{18} = a_6 - a_7$ $b_{10} = (-5h_0 - 5h_1 + 5h_2 + 3h_3 + 5h_4 - 5h_5 - 5h_6 + h_7)/20$
$a_{19} = a_{17} + a_{18}$ $b_{11} = (5h_1 - 2h_3 - 5h_4 + 5h_6 + h_7)/20$
$a_{20} = x_0 + x_4$ $b_{12} = h_0 - h_3 + h_4$
$a_{21} = x_0$ $b_{13} = -2h_0 + h_3 - h_7$
$a_{22} = x_4$ $b_{14} = h_3 - 2h_4 + h_7$
$a_{23} = x_3 + x_7$ $b_{15} = -h_2 + h_6 - 2h_7$
$a_{24} = x_3$ $b_{16} = 2h_3 - 2h_6 + 2h_7$
$a_{25} = x_7$ $b_{17} = 2h_2 - 2h_3 + 2h_7$
$a_{26} = a_{15} - a_9 + x_0 - a_{23}$ $b_{18} = (-h_3 + h_7)/5$
$a_{27} = a_8 - a_{14} + x_3 + x_4 - x_7$ $b_{19} = (-h_3 - h_7)/5$
$a_{28} = a_{26} + a_{27}$ $b_{20} = h_3/5$

$$m_k = a_{k+8} b_k \qquad k = 0, \dots, 20$$

$u_0 = m_0 + m_2$ $u_{10} = u_9 + m_{13}$ $u_{20} = u_9 + m_{14}$
$u_1 = m_1 + m_2$ $u_{11} = u_0 + u_2$ $u_{21} = u_{19} + u_{19}$
$u_2 = m_3 + m_5$ $u_{12} = u_0 - u_2$ $u_{22} = m_{15} - u_{21}$
$u_3 = m_4 + m_5$ $u_{13} = u_1 + u_3$ $u_{23} = u_{22} + m_{16}$
$u_4 = m_6 + m_8$ $u_{14} = u_1 - u_3$ $u_{24} = -u_{22} - m_{17}$
$u_5 = m_7 + m_8$ $u_{15} = u_4 + u_6$ $u_{25} = u_8 + u_8$
$u_6 = m_{11} + m_9$ $u_{16} = u_4 - u_6$ $u_{26} = u_{25} + u_{25}$
$u_7 = m_{10} + m_{11}$ $u_{17} = u_5 + u_7$ $u_{27} = u_{19} + u_{25}$
$u_8 = m_{20} + m_{19}$ $u_{18} = u_5 - u_7$
$u_9 = m_{12} + u_8$ $u_{19} = m_{18} - m_{19}$

$$y_0 = u_{13} + u_{17} + u_{20}$$
$$y_1 = u_{12} - u_{18} + u_{23}$$
$$y_2 = u_{11} - u_{15} + u_{25}$$
$$y_3 = u_{12} + u_{18} + u_{27}$$

$$y_4 = u_{11} + u_{15} + u_{10} + u_{19}$$
$$y_5 = -u_{14} - u_{16} + u_{24} + u_{26}$$
$$y_6 = -u_{13} + u_{17} + u_{21} + u_{25}$$
$$y_7 = -u_{14} + u_{16} + u_{19}$$

3.6 Appendix – Reduced DFT Algorithms for $N=4, 8, 9, 16$

These algorithms compute $q^{t-1}(q-1)$ output samples of DFTs of dimension $N = q^t$, with q prime, where the last q^{t-1} input samples $\{x_n\}$ are null:

$$\bar{X}_k = \sum_{n=0}^{q^{t-1}(q-1)-1} x_n W^{nk} \qquad 1 \le k \le N-1$$

$$k \not\equiv 0 \text{ modulo } q$$

$$W = \exp -j2\pi/N, j = \sqrt{-1}.$$

The algorithms are arranged in the same format as in Sect. 3.5. Trivial multiplications by ± 1, $\pm j$ are given between parentheses.

3.6.1 $N=4$

2 Multiplications (2), 2 Additions

$$m_0 = 1 \cdot x_0$$
$$m_1 = -j \cdot x_1$$
$$\bar{X}_1 = m_0 + m_1$$
$$\bar{X}_3 = m_0 - m_1$$

3.6.2 $N=8$ $u=\pi/4$

4 Multiplications (2), 10 Additions

$$m_0 = 1 \cdot x_0 \qquad m_1 = (x_1 - x_3) \cos u$$
$$m_2 = -jx_2 \qquad m_3 = -j(x_1 + x_3) \sin u$$

$$s_1 = m_0 + m_1 \qquad s_2 = m_0 - m_1$$
$$s_3 = m_2 + m_3 \qquad s_4 = m_2 - m_3$$

$$\bar{X}_1 = s_1 + s_3 \qquad \bar{X}_3 = s_2 - s_4 \qquad \bar{X}_5 = s_2 + s_4 \qquad \bar{X}_7 = s_1 - s_3$$

3.6.3 $N=16$ $u=2\pi/16$

10 multiplications (2), 32 Additions

$$t_1=x_1+x_7 \qquad t_2=x_1-x_7 \qquad t_3=x_3+x_5 \qquad t_4=x_5-x_3$$

$$m_0=1\cdot x_0 \qquad m_1=(x_2-x_6)\cos 2u \qquad\qquad m_2=(t_2+t_4)\cos 3u$$
$$m_3=(\cos u+\cos 3u)t_2 \qquad\qquad\qquad m_4=(\cos 3u-\cos u)t_4$$
$$m_5=-j\cdot x_4 \qquad m_6=-j(x_2+x_6)\sin 2u \qquad m_7=-j(t_1+t_3)\sin 3u$$
$$m_8=j(\sin 3u-\sin u)\cdot t_1 \qquad\qquad\qquad m_9=-j(\sin u+\sin 3u)\cdot t_3$$

$$
\begin{array}{lll}
s_1=m_0+m_1 & s_2=m_0-m_1 & s_3=m_3-m_2 \\
s_4=m_4-m_2 & s_5=s_1+s_3 & s_6=s_1-s_3 \\
s_7=s_2+s_4 & s_8=s_2-s_4 & s_9=m_5+m_6 \\
s_{10}=m_5-m_6 & s_{11}=m_7+m_8 & s_{12}=m_7-m_9 \\
s_{13}=s_9+s_{11} & s_{14}=s_9-s_{11} & s_{15}=s_{10}+s_{12} \\
s_{16}=s_{10}-s_{12} & &
\end{array}
$$

$$
\begin{array}{lll}
\bar{X}_1=s_5+s_{13} & \bar{X}_3=s_8-s_{16} & \bar{X}_5=s_7+s_{15} \\
\bar{X}_7=s_6-s_{14} & \bar{X}_9=s_6+s_{14} & \bar{X}_{11}=s_7-s_{15} \\
\bar{X}_{13}=s_8+s_{16} & \bar{X}_{15}=s_5-s_{13} &
\end{array}
$$

3.6.4 $N=9$ $u=2\pi/9$

8 Multiplications, 28 Additions

$$t_1=x_4+x_5 \qquad t_2=x_4-x_5$$
$$m_0=\tfrac{1}{2}(x_0+x_0-x_3)$$
$$m_1=\left(\frac{2\cos u-\cos 2u-\cos 4u}{3}\right)\cdot(x_1-x_2)$$
$$m_2=\left(\frac{\cos u+\cos 2u-2\cos 4u}{3}\right)\cdot(x_2-t_1)$$
$$m_3=\left(\frac{\cos u-2\cos 2u+\cos 4u}{3}\right)\cdot(t_1-x_1)$$
$$m_4=-j(\sin 3u)\cdot x_3$$
$$m_5=-j(\sin u)\cdot(x_1+x_2)$$
$$m_6=-j(\sin 4u)\cdot(x_2+t_2)$$
$$m_7=j(\sin 2u)\cdot(x_1-t_2)$$

$$
\begin{array}{ll}
s_2=m_1+m_2+m_0 & s_3=-m_2+m_3+m_0 \\
s_4=-m_1-m_3+m_0 & s_5=m_4+m_5+m_6 \\
s_6=-m_6+m_7+m_4 & s_7=-m_5-m_7+m_4
\end{array}
$$

$$
\begin{array}{lll}
\bar{X}_1=s_2+s_5 & \bar{X}_2=s_3-s_6 & \bar{X}_4=s_4+s_7 \\
\bar{X}_5=s_4-s_7 & \bar{X}_7=s_3+s_6 & \bar{X}_8=s_2-s_5
\end{array}
$$

References

3.1 S. Winograd: Math. Comput. **32**, 175–199 (1978)
3.2 R. C. Agarwal, J. W. Cooley: IEEE Trans. ASSP-**25**, 392–410 (1977)
3.3 H. J. Nussbaumer: Electron. Lett. **13**, 386–387 (1977)
3.4 H. J. Nussbaumer: "New Algorithms for Convolution and DFT Based on Polynomial Transforms", in IEEE 1978 Intern. Conf. Acoust., Speech, Signal Processing Proc., pp. 638–641
3.5 H. J. Nussbaumer, P. Quandalle: IBM J. Res. Dev. **22**, 134–144 (1978)
3.6 D. J. Winter: *The Structure of Fields* (Springer, Berlin, Heidelberg, New York 1974)
3.7 T. Nagell: *Introduction to Number Theory* (Chelsea, New York 1964)
3.8 C. M. Rader: IEEE Trans. C-**21**, 1269–1273 (1972)
3.9 R. C. Agarwal, C. S. Burrus: Proc. IEEE **63**, 550–560 (1975)
3.10 B. Gold, C. M. Rader, A. V. Oppenheim, T. G. Stockham: *Digital Processing of Signals*, (Mc Graw Hill, New York 1969) Ch. 7, pp. 203–213
3.11 S. Winograd: "Some Bilinear Forms Whose Multiplicative Complexity Depends on the Field of Constants"; IBM Res. Rpt. RC5669, IBM Watson Research Center, Yorktown Heights, N. Y. (1975)
3.12 J. W. Cooley, J. W. Tukey: Math. Comput. **19**, 297–301 (1965)
3.13 P. Quandalle: "Filtrage numérique rapide par transformées de Fourier et transformées polynômiales. Etude de l'implantation des algorithmes sur microprocesseurs"; Ph. D. Thesis, University of Nice, France (1979)
3.14 R. C. Agarwal, J. W. Cooley: "New Algorithms for Digital Convolution", in 1977 Intern. Conf., Acoust., Speech and Signal Processing Proc., p. 360
3.15 C. M. Rader, N. M. Brenner: IEEE Trans. ASSP-**24**, 264–266 (1976)
3.16 C. S. Burrus: "Digital Filter Realization by Distributed Arithmetic", in Proc. 1976 IEEE Intern. Symp. Circuits and Systems, Munich (1976) pp. 106–109
3.17 H. J. Nussbaumer, P. Quandalle: IEEE Trans. ASSP-**27**, 169–181 (1979)
3.18 A. V. Oppenheim, R. W. Schafer: *Digital Signal Processing*, (Prentice Hall, Englewood Cliffs, N. J. 1975) pp. 320–321
3.19 C. M. Rader: Proc. IEEE **56**, 1107–1108 (1968)
3.20 I. J. Good: IEEE Trans. C-**20**, 310–317 (1971)
3.21 D. P. Kolba, T. W. Parks: IEEE Trans. ASSP-**25**, 281–294 (1977)
3.22 R. C. Agarwal, C. S. Burrus: IEEE Trans. ASSP-**22**, 87–97 (1974)

4. Winograd's Discrete Fourier Transform Algorithm

S. Zohar

With 21 Figures

The new DFT algorithm of S. Winograd is developed and presented in detail. This is an algorithm which uses about 1/5 of the number of multiplications used by the Cooley-Tukey algorithm and is applicable to any order which is a product of relatively prime factors from the following list: 2, 3, 4, 5, 7, 8, 9, 16. The algorithm is presented in terms of a series of tableaus – one for each term in this list – which are convenient, compact, graphical representations of the sequence of arithmetic operations in the corresponding parts of the algorithm. Using these in conjunction with Tables 4.5, 6, makes it relatively easy to apply the algorithm and evaluate its performance.

The development of the subject is organized in a way which allows extensive skipping on a first reading.

4.1 An Overview

Ever since the discovery of the FFT algorithm [4.1], the following question must have been phrased in many minds: "Does the FFT algorithm represent the ultimate in the fast computation of the discrete Fourier transform (DFT) or is there a still faster algorithm yet to be discovered?" One answer was provided in 1968 by *Yavne* [4.2] who showed that the number of multiplications could be halved while leaving the number of additions unchanged. More recently (1976), a significant step along this path was taken by *Winograd* [4.3] who developed an algorithm which reduces the number of multiplications of the radix-2 FFT algorithm [4.1] by a factor of about 5. This reduction is accompanied by a small increase or decrease in the number of additions. In most cases, the increase does not exceed 20%.

Our basis for comparison both here and later on is the "nominal" performance of the Cooley-Tukey (FFT) algorithm, namely, the computation of an Nth order DFT of complex data with $\mathcal{M}_{CT}$ real multiplications and $\mathcal{A}_{CT}$ real additions where[1]

$$\mathcal{M}_{CT} = 2N \log_2 N \; ; \quad \mathcal{A}_{CT} = 1.5 \mathcal{M}_{CT} . \tag{4.1}$$

We adopt (4.1) as the basis for comparison for all N.

---

1 Equation (4.1) is adopted as a convenient yardstick. It should be borne in mind that in addition to Yavne's algorithm, there are other FFT variants which are somewhat more efficient than (4.1).

Winograd's algorithm then performs the above task using about $\mathcal{M}_{CT}/5$ real multiplications. We devote the rest of this section to a description of the capabilities and constraints of the algorithm so that the reader could assess its suitability to his needs before delving any deeper.

At its present state of development, the algorithm is applicable to any N satisfying

$$N = \prod_{k=1}^{\kappa} N_k \tag{4.2}$$

in which the N_k's are relatively prime factors taken from the following list

$$N_k = 2, 3, 4, 5, 7, 8, 9, 16. \tag{4.3}$$

The maximal N is therefore $16 \cdot 9 \cdot 7 \cdot 5 = 5040$. All of the N values satisfying the above prescriptions are listed in the summary of the algorithm presented in Table 4.6 (p. 154). The actual multiplication reduction factor is listed there for each N, under the heading G_∞. Note that the average of G_∞ for all $N > 140$ is about 5.5. With such a large reduction in the number of multiplications, it is reasonable to expect that the new algorithm will run faster than the Cooley-Tukey algorithm in most systems. To make a more specific claim, we must know the basic system parameter μ which is the ratio of the time taken to execute one real multiplication to the time taken to execute one real addition. (The term "real" is used here as opposed to "complex"; not as opposed to "integer" in the Fortran language.) For very large μ (microprocessors, software multipliers, etc.), the speed gain approaches G_∞ asymptotically. For lower values of μ, the gain would be smaller. Denoting the speed gain by G, it will be shown (Sect. 4.8) that

$$G(\mu) = G_\infty \frac{\mu + 1.5}{\mu + R} \tag{4.4}$$

where R is another parameter listed with G_∞ in Table 4.6. Obviously, it is now a trivial matter to compute the speed gain for any system and any permissible N.

It should be pointed out that (4.4) is based on the time for arithmetic operations only and that the structural complexity of Winograd's algorithm will tend to slow down its software implementations.

The main disadvantage of the new algorithm is its need for a large memory. We express this in terms of the parameter $\mathcal{M}$ of Table 4.6. For the processing of complex data we have to have a storage array of $1.5\,\mathcal{M}$ real words. $\mathcal{M}$ varies from about $2N$ at the lower end of the table to about $4N$ at its upper end. Thus, for high N values, we require a storage array of size $6N$. This is $4N$ more than the minimal requirements of $2N$ for storing the input vector.

Of the total memory requirement of $1.5\,\mathcal{M}$ real words, $0.5\,\mathcal{M}$ are needed for the storage of precomputed constants. Since not all of these constants are distinct, it is probably feasible to reduce this part by the use of a more involved addressing scheme.

Another probable disadvantage of the new algorithm is that, in comparison with the Cooley-Tukey algorithm, it might require more bits per word to maintain a prescribed level of precision. This effect is discussed in some more detail in Sect. 4.9 but the whole subject merits further study.

The development of the algorithm consists of two distinct parts. In the first part, fast DFT algorithms for the low orders listed in (4.3) are developed as a set of building blocks. The second part introduces a combining algorithm which integrates groups of these building blocks into the desired final structures, namely, DFT algorithms for orders N prescribed by (4.2).

The low-order algorithms of (4.3) are derivable from algorithms of orders 2, 4, 6 of another type of transformation called LCT (left-circulant transformation, more commonly referred to as circular correlation; see Sect. 4.2). Hence, the following structure of the paper: Section 4.2 is devoted to the DFT-LCT interrelationship. This is followed by the three LCT algorithms in Sect. 4.3 and the seven DFT algorithms derived from them in Sects. 4.4–6. Section 4.7 tackles the integration of these low-order algorithms into the desired algorithm for order N satisfying (4.2). Section 4.8 is devoted to performance evaluation and Sect. 4.9 concludes with an overview and some comments regarding "in-place" transformation.

The treatment of the subject is detailed and relatively complete, aiming to provide a sound basis for further development of the subject. Naturally, this demands a substantial investment of time. It should be pointed out, however, that a reasonable grasp of the basic ideas and their application can be obtained by skipping the detailed derivations of the low-order algorithms. If this is acceptable, the following parts of the paper may prove sufficient: Section 4.2, first part of Sect. 4.3 [up to the treatment of the left circulant of order 4 (Sect. 4.3.2)], the introduction of the η vector on p. 119, the tableau generalization in Sect. 4.6 [portion bounded by (4.126, 127)], Sect. 4.7, last part of Sect. 4.8 [following (4.181)], and Sect. 4.9.

4.2 Strategy Development

The cornerstone of Winograd's algorithm is a theorem [4.4] providing the solution to a seemingly unrelated problem: Given the two polynomials $A(x)$, $B(x)$, what is the minimal number of multiplications required to compute

$$\{A(x)B(x)\} \bmod C(x), \tag{4.5}$$

where $C(x)$ is a given monic polynomial. The connection with the DFT consists of two links: Firstly, the DFT matrix is shown to be related to another special transformation in which the transforming matrix is a left circulant (exact definition follows later). Secondly, the evaluation of this transformation is shown to be identical with the evaluation of (4.5). Thus, the minimization of the number of multiplications in (4.5) leads via the above two links to a minimization of the number of multiplications in the computation of the DFT.

We proceed now with some required definitions: A Hankel matrix is one in which the value of element a_{ij} is a function of $(i+j)$. In such a matrix, one encounters identical elements as one moves along any diagonal sloping down and to the left. Obviously, the matrix is completely determined by its first row and last column.

The matrix we shall be concerned with here is a special case of a Hankel matrix, namely, a Hankel matrix for which (for order n and index range $0, 1, \ldots, n-1$)

$$a_{\varrho,n-1}=a_{0,\varrho-1} \quad (1\leq\varrho\leq n-1), \tag{4.6}$$

that is, the last column is a trivial rearrangement of the elements of the first row. Hence, this matrix is completely prescribed by its first row. Indeed, the second row is obtained from the first one by a circular left shift (element shifted out on the left, reappears on the right), the third is derived the same way from the second, and so on. We call such a matrix a left circulant[2] or, equivalently, an LC matrix. Similarly, the linear transformation effected by such a matrix will be referred to as an LCT (left-circulant transformation).

Equation (4.7) illustrates the general LCT of order 3

$$\begin{bmatrix} c_0 \\ c_1 \\ c_2 \end{bmatrix} = \begin{bmatrix} a_2 & a_1 & a_0 \\ a_1 & a_0 & a_2 \\ a_0 & a_2 & a_1 \end{bmatrix} \begin{bmatrix} b_0 \\ b_1 \\ b_2 \end{bmatrix}. \tag{4.7}$$

Note that (4.7) can be regarded as a prescription for getting the sequence (c_0, c_1, c_2) from the sequences (a_0, a_1, a_2) and (b_0, b_1, b_2). From this point of view, $\{c_i\}$ is referred to as the circular correlation of the sequences $\{a_i\}, \{b_i\}$. Alternatively, the reversed sequence (c_2, c_1, c_0) is referred to as the circular convolution of the sequences $\{a_i\}, \{b_i\}$.

Many of the subsequent mathematical manipulations involve a matrix which, while not being an LC itself, does contain an LC submatrix. We refer to such a matrix as a quasi-left-circular matrix (QLC).

We turn now to the LCT-DFT link. As will become obvious later on, we should concern ourselves with a trivial modification of the DFT defined as follows:[*]

$$W=\exp\left(-i\frac{2\pi}{N}\right) \tag{4.8}$$

$$F_u=\Omega \sum_{v=0}^{N-1} W^{uv}f_v \quad (u=0, 1, \ldots, N-1). \tag{4.9}$$

---

2 This is based on the term circulant which is commonly used to describe a matrix generated from its first row by circular right shifts.

* The expression i for $\sqrt{-1}$ in this chapter corresponds to the expression j in the other chapters.

Ω in (4.9) is an arbitrary complex constant. When $\Omega=1$, (4.9) reduces to the standard DFT.

Let N, the order of the DFT, satisfy

$$
\left.
\begin{array}{l}
N=p^k \quad (p \text{ prime}; k \text{ integer}) \\[2mm]
k < \begin{cases} \infty & (p \text{ odd}) \\ 3 & (p=2) \end{cases}
\end{array}
\right\}.
\tag{4.10}
$$

We proceed to show now that for such N, (4.9) can be brought into the form of a QLC matrix[3]. The derivation follows [4.5] and is based on the number-theoretic idea of a primitive root: g, the primitive root of N [satisfying (4.10)] is an integer whose integral powers (mod N) generate all integers in the interval $(1, N)$ except multiples of p.

The number of multiples of p in the above interval is

$$
\frac{N}{p}=p^{k-1}.
\tag{4.11}
$$

Therefore, the number of integers generated by g is

$$
n=p^k-p^{k-1}=(p-1)p^{k-1}
\tag{4.12}
$$

and we may say that the sequence

$$
\{g^\varrho \bmod N\} \quad (\varrho=0,1,\ldots,n-1)
\tag{4.13}
$$

is just a permutation of those integers in the interval $(1, N)$ which are not multiples of p.

We use these ideas now to relabel the indices of (4.9) as follows. All indices which are not multiples of p will be represented as in (4.13). Specifically, denoting

$$
\left.
\begin{array}{l}
r=g^\varrho \bmod N \\
s=g^\sigma \bmod N
\end{array}
\right\} \quad (\varrho, \sigma=0,1,\ldots,n-1),
\tag{4.14}
$$

we define

$$
B_\varrho=F_r ; \quad b_\sigma=f_s \quad (\varrho, \sigma=0,1,\ldots,n-1).
\tag{4.15}
$$

---

3 This is true for a range wider than (4.10). However, (4.10) is sufficient for our purpose.

The indices which are multiples of p are now used to define

$$B_{(i)} = F_i ; \quad b_{(i)} = f_i \quad (i \bmod p = 0). \tag{4.16}$$

The introduction of the parenthesized index (i) merits some elaboration. What we are considering here is some sort of scrambling (relabeling) of the entities F_r (and f_s). From what we have seen so far, we can subdivide the N terms comprising the set $\{F_r\}$ into 2 subsets: the n-term set in which r is expressible by (4.14) and the remainder set of $N-n$ terms in which r is a multiple of p. Using the definition (4.15), we can denote the terms of the first subset $B_0, B_1, \ldots, B_{n-1}$. Similarly, we could define $B_{n+m} = F_{mp}$ (m integer) which would conveniently label the terms of the second set $B_n, B_{n+1}, \ldots, B_{N-1}$. There is, however, no need to do that and this approach becomes cumbersome in Sect. 4.6 where $\{F_r\}$ has to be subdivided into 3 subsets. The approach adopted here is based on the fact that the indexing of the terms of the second subset is quite arbitrary provided that we make sure that the set referred to is clearly identified. As is evident in (4.16), we chose to index the terms of the second subset by simply copying the index of the corresponding F_r terms. The index parentheses in $B_{(i)}$ is just a means of identifying this term as a member of the second subset. Note in this context that the index zero appears in both subsets. Thus,

$$B_0 = F_1 ; \quad B_{(0)} = F_0 . \tag{4.17}$$

Having introduced the required terminology, we embark now on the elimination of F_u, f_v from (4.9). In doing this, we split the summation of (4.9) into two parts. In the first part $v = mp$ (m integer), that is, $v \bmod p = 0$. In the second part $v = s$. The result is[4]

$$B_{(tp)} = F_{tp} = \Omega \sum_{m=0}^{N-n-1} W^{mtp^2} b_{(mp)} + \Omega \sum_{\sigma=0}^{n-1} W^{tpg\sigma} b_\sigma \quad (t=0,1,\ldots,N-n-1) \tag{4.18}$$

$$B_\varrho = \hat{B}_\varrho + B'_\varrho \quad (\varrho = 0, 1, \ldots, n-1), \tag{4.19}$$

where

$$\hat{B}_\varrho = \Omega \sum_{m=0}^{N-n-1} W^{mpg\varrho} b_{(mp)} \quad (\varrho = 0, 1, \ldots, n-1) \tag{4.20}$$

$$B'_\varrho = \Omega \sum_{\sigma=0}^{n-1} W^{(g\varrho+\sigma)} b_\sigma \quad (\varrho = 0, 1, \ldots, n-1). \tag{4.21}$$

---

4 We take here advantage of the fact that (4.8) ensures $W^{(m \bmod N)} = W^m$.

The term $(\varrho + \sigma)$ identifies (4.21) as a Hankel transformation, that is, if we write down (4.21) with ϱ and σ as the row and column indices, respectively, then the matrix transforming b into B' is a Hankel matrix. More than that, it is that special kind of a Hankel matrix referred to earlier as a left circulant. To see this, note that in view of the LC condition (4.6), our Hankel matrix would be an LC if

$$g^{\varrho+n-1} = g^{\varrho-1} \bmod N . \tag{4.22}$$

But since the primitive root of N always satisfies[5]

$$g^n = 1 \bmod N , \tag{4.23}$$

it is obvious that (4.22) is indeed true and (4.21) is an LCT. We conclude that the permutations (4.14–16) will transform any DFT matrix of order N prescribed by (4.10), into a QLC matrix whose LC portion is of order n.

Of particular interest is the special case in which N is prime, that is, $k = 1$; $N = p$, so that (4.12) yields

$$n = N - 1 \tag{4.24}$$

and the sequence (4.13) is just a permutation of the integers $1, 2, \ldots, N-1$. In this case, (4.18, 20) simplify as follows:

$$B_{(0)} = \Omega \left(b_{(0)} + \sum_{\sigma=0}^{n-1} b_\sigma \right) \tag{4.25}$$

$$\hat{B}_\varrho = \Omega b_{(0)} \qquad (\varrho = 0, 1, \ldots, n-1). \tag{4.26}$$

To illustrate these ideas, consider the case of $N = 7$. The least positive primitive root of 7 is 3 [4.6]. Indeed, direct computation yields

ϱ	0	1	2	3	4	5
$3^\varrho \bmod 7$	1	3	2	6	4	5

(4.27)

---

5 In the standard treatment of primitive roots, (4.13) is usually replaced by

$$\{g^\varrho \bmod N\} \qquad (\varrho = 1, 2, \ldots, n). \tag{4.13'}$$

Congruence (4.23) shows that the two formulations are equivalent. To prove (4.23) assume $g^m = 1 \bmod N$ for $m < n$. It follows then that $g^{m+1} = g^1 \bmod N$ and thus two of the terms in sequence (4.13') are equal. Hence the contradiction that g is not a primitive root. Conclusion: $m = n$.

Applying (4.14–16), we get the following form:

$$
\begin{bmatrix} F_0 \\ F_1 \\ F_3 \\ F_2 \\ F_6 \\ F_4 \\ F_5 \end{bmatrix}
\begin{bmatrix} B_{(0)} \\ B_0 \\ B_1 \\ B_2 \\ B_3 \\ B_4 \\ B_5 \end{bmatrix}
= \Omega
\left[\begin{array}{c|cccccc}
W^0 & W^0 & W^0 & W^0 & W^0 & W^0 & W^0 \\
W^0 & W^1 & W^3 & W^2 & W^6 & W^4 & W^5 \\
W^0 & W^3 & W^2 & W^6 & W^4 & W^5 & W^1 \\
W^0 & W^2 & W^6 & W^4 & W^5 & W^1 & W^3 \\
W^0 & W^6 & W^4 & W^5 & W^1 & W^3 & W^2 \\
W^0 & W^4 & W^5 & W^1 & W^3 & W^2 & W^6 \\
W^0 & W^5 & W^1 & W^3 & W^2 & W^6 & W^4
\end{array}\right]
\begin{bmatrix} b_{(0)} \\ b_0 \\ b_1 \\ b_2 \\ b_3 \\ b_4 \\ b_5 \end{bmatrix} ;
\quad
\begin{bmatrix} b_{(0)} \\ b_0 \\ b_1 \\ b_2 \\ b_3 \\ b_4 \\ b_5 \end{bmatrix}
=
\begin{bmatrix} f_0 \\ f_1 \\ f_3 \\ f_2 \\ f_6 \\ f_4 \\ f_5 \end{bmatrix}.
\qquad (4.28)
$$

The LC structure is quite apparent here.

So far, we have established the first link to problem (4.5) for N satisfying (4.10). This constraint on N will be relaxed later on. We turn now to the second link, namely, showing that evaluation of an LCT is equivalent to evaluation of (4.5). We intend to establish this equivalence and, from it, derive algorithms for some general low-order LC transformations. It should be pointed out, however, that the LC matrix we are concerned with is one obtained by permuting a DFT submatrix and, as such, is still a function of only one variable (W) whereas the general LC matrix of order n is a function of n variables. This suggests further simplifications in our case which will indeed be realized later on.

The matrix multiplication we are considering is shown in (4.29) where the LC pattern is clearly visible.

$$
\begin{bmatrix} t_m \\ t_{m-1} \\ t_{m-2} \\ \vdots \\ t_2 \\ t_1 \\ t_0 \end{bmatrix}
\begin{bmatrix}
a_m & a_{m-1} & a_{m-2} & \cdots & a_2 & a_1 & a_0 \\
a_{m-1} & a_{m-2} & & & & a_0 & a_m \\
a_{m-2} & & & & & & a_{m-1} \\
\vdots & & & & & & \vdots \\
a_2 & & & & & & a_3 \\
a_1 & & & & & a_3 & a_2 \\
a_0 & a_m & a_{m-1} & \cdots & a_3 & a_2 & a_1
\end{bmatrix}
\begin{bmatrix} b_0 \\ b_1 \\ b_2 \\ \vdots \\ b_{m-2} \\ b_{m-1} \\ b_m \end{bmatrix}.
\qquad (4.29)
$$

We introduce now the auxiliary polynomials. (*Note:* The polynomial subscript indicates its degree.)

$$
A_m(x) = \sum_{i=0}^{m} a_i x^i \qquad (4.30)
$$

$$
B_m(x) = \sum_{i=0}^{m} b_i x^i \qquad (4.31)
$$

$$
T_m(x) = \sum_{i=0}^{m} t_i x^i \qquad (4.32)
$$

Consider the product polynomial

$$V_{2m}(x) = A_m(x)B_m(x) = \sum_{i=0}^{2m} v_i x^i. \tag{4.33}$$

It is easy to see that the coefficients of this polynomial are obtainable by the matrix product (4.34).

$$
\begin{bmatrix}
v_{2m} \\
\cdot \\
\cdot \\
\cdot \\
\cdot \\
\cdot \\
v_{m+1} \\
v_m \\
v_{m-1} \\
\cdot \\
\cdot \\
\cdot \\
\cdot \\
\cdot \\
v_0
\end{bmatrix}
=
\begin{bmatrix}
0 & 0 & 0 & \cdots & 0 & 0 & a_m \\
0 & 0 & & & & a_m & a_{m-1} \\
0 & & & & & & a_{m-2} \\
\vdots & & & & & & \vdots \\
0 & & & & & & a_2 \\
0 & & & & & a_2 & a_1 \\
a_m & a_{m-1} & a_{m-2} & \cdots & a_2 & a_1 & a_0 \\
a_{m-1} & a_{m-2} & & & & a_0 & 0 \\
a_{m-2} & & & & & & 0 \\
\vdots & & & & & & \vdots \\
a_2 & & & & & & 0 \\
a_1 & a_0 & & & & 0 & 0 \\
a_0 & 0 & 0 & \cdots & 0 & 0 & 0
\end{bmatrix}
\begin{bmatrix}
b_0 \\
b_1 \\
b_2 \\
\vdots \\
b_{m-2} \\
b_{m-1} \\
b_m
\end{bmatrix}
. \tag{4.34}
$$

Comparing this to (4.29), we see that interchanging the two indicated triangular sections in (4.34) will transform the matrix of (4.34) into a trivial augmentation of the matrix of (4.29). This fact can be translated to the following relationship between $T_m(x)$ and $V_{2m}(x)$:

$$T_m(x) = V_{2m}(x) - (a_m b_1 + a_{m-1} b_2 + \ldots + a_1 b_m)(x^{m+1} - 1)$$
$$- (a_m b_2 + \ldots + a_2 b_m)x(x^{m+1} - 1)$$
$$\vdots$$
$$- (a_m b_{m-1} + a_{m-1} b_m)x^{m-2}(x^{m+1} - 1)$$
$$- a_m b_m x^{m-1}(x^{m+1} - 1) \tag{4.35}$$
$$\therefore\ T_m(x) = V_{2m}(x) - (x^{m+1} - 1)F_{m-1}(x), \tag{4.36}$$

where $F_{m-1}(x)$ is the indicated polynomial of degree $(m-1)$. Hence

$$\frac{V_{2m}(x)}{x^{m+1} - 1} = F_{m-1}(x) + \frac{T_m(x)}{x^{m+1} - 1}. \tag{4.37}$$

Note that, in the quotient on the right, the denominator degree is higher than the numerator degree. This means that $T_m(x)$ is just the remainder obtained when dividing $V_{2m}(x)$ by $(x^{m+1}-1)$. In other words,

$$T_m(x) = \{A_m(x)B_m(x)\} \bmod (x^n - 1). \tag{4.38}$$

We have made use here of the fact that the order of the LC matrix in (4.29) is

$$n = m + 1. \tag{4.39}$$

Equation (4.38) is a prescription for performing the LCT of (4.29) through polynomial manipulations identical with those of (4.5). This, then, establishes the second link.

Consider now the number of multiplications required to evaluate (4.29). Straight matrix multiplication requires n^2 scalar multiplications. A much lower value is prescribed by Winograd's theorem [4.4]. In its narrower application to the present case, the theorem states that if $(x^n - 1)$ is representable as

$$x^n - 1 = \prod_{i=1}^{k(n)} m_i(x), \tag{4.40}$$

where the $m_i(x)$ are distinct polynomials irreducible over the field of rationals, then the minimal number of multiplications is $(2n - k)$, provided that multiplications by rational numbers are not counted.

The exclusion of rational multiplications merits an explanation. Suppose we have a minimal realization of $T_m(x)$ of the following form:

$$T_m(x) = \sum_{r=1}^{R} \left(\frac{J_r}{K_r}\right) F_r(x), \tag{4.41}$$

where J_r, K_r are integers and in which $F_r(x)$ involves no rational multiplications. According to the theorem, the F_r's will require a total of $(2n - k)$ multiplications and we are essentially being told that the additional R rational multiplications appearing in (4.41) do not count. To see what is involved here, let us clear fractions in (4.41). Let K be the least common denominator and let

$$\frac{J_r}{K_r} = \frac{J'_r}{K}. \tag{4.42}$$

Hence,

$$K T_m(x) = \sum_{r=1}^{R} J'_r F_r(x). \tag{4.43}$$

Each of the multiplications by J'_r can be implemented as $(J'_r - 1)$ additions so that $KT_m(x)$ does not require any multiplications above the $(2n - k)$ used in computing the F_r's. Finally, we compensate for the multiplication by K on the left, by prescaling the a matrix, that is, replacing a_i, $A_m(x)$ by

$$\hat{a}_i = \frac{a_i}{K}; \quad \hat{A}_m(x) = \sum_{i=0}^{m} \hat{a}_i x^i. \qquad (4.44)$$

Thus, (4.38) is now replaced by

$$T_m(x) = K\{\hat{A}_m(x)B_m(x)\} \bmod (x^n - 1), \qquad (4.45)$$

and we see that multiplications by rationals can always be eliminated without an increase in the number of irrational multiplications.

It should be pointed out that while this argument is theoretically sound, practically, one should be concerned with the cost in terms of the extra additions introduced to eliminate the rational multiplications. Obviously, when these extra additions take longer than the multiplications they replace, we would be better off leaving the multiplications in. This will, indeed, be the case when n is large. Therefore, the algorithm taking advantage of Winograd's theorem has to be constructed in such a way that a DFT of large order is broken down into many LC transformations of low order. As a matter of fact, all the DFT orders appearing in Table 4.6 ($N_{max} = 5040$) call for just three LC transformations of orders 2, 4, 6.

The factoring of $(x^n - 1)$ for these three cases is shown in Table 4.1 in which the last column gives the minimal number of multiplications as stated by Winograd's theorem. In the next section we develop the specific algorithms which realize these minima. These three algorithms serve as a foundation for the subsequent construction of DFT algorithms for all orders listed in (4.3).

4.3 The Basic LCT Algorithms

Our approach here is to present the general method first in sufficient detail so that its application to the three specific n values can be subsequently presented as a mostly self-explanatory sequence of equations.

The starting point is (4.45) in which K is left indeterminate till the very end of the derivation. The factoring of $(x^n - 1)$ is spelled out in Table 4.1 which

Table 4.1. Rational factorization of $(x^n - 1)$

n	$x^n - 1$	k	$2n - k$
2	$(x-1)(x+1)$	2	2
4	$(x-1)(x+1)(x^2+1)$	3	5
6	$(x-1)(x+1)(x^2+x+1)(x^2-x+1)$	4	8

identifies the $m_i(x)$ factors of (4.40). With the m_i's available, we evaluate $T_m(x)$ in a two-phase scheme based on the polynomial version of the Chinese Remainder Theorem [4.7]. In phase 1 we compute

$$u_i(x) = \{\hat{A}_m(x)B_m(x)\} \bmod m_i(x) \tag{4.46}$$

for all i of (4.40). This is based entirely on results established in the Appendix and summarized in Tables A.1, 2 there. In phase 2 we use the polynomial version of Garner's algorithm [4.7] to construct $T_m(x)$ from the u_i's. This calls for the auxiliary functions $c_{ij}(x)$, $v_i(x)$ introduced below. Their utilization in the construction of $T_m(x)$ is spelled out in (4.50).

$$[m_i(x)c_{ij}(x)] \bmod m_j(x) = 1 \quad \text{[definition of } c_{ij}(x)] \tag{4.47}$$

$$v_1(x) = u_1(x) \tag{4.48}$$

$$v_i(x) = \{[\ldots[[[u_i(x) - v_1(x)]c_{1,i}(x) - v_2(x)]c_{2,i}(x) - v_3(x)]$$
$$\cdot c_{3,i}(x) -, \ldots, - v_{i-1}(x)]c_{i-1,i}(x)\} \bmod m_i(x) \tag{4.49}$$

$$T_m(x) = K\left[v_1(x) + \sum_{i=2}^{k(m+1)} v_i(x) \prod_{j=1}^{i-1} m_j(x)\right] \quad (k \text{ from Table 4.1}). \tag{4.50}$$

The computation of $c_{ij}(x)$ is trivial when $m_j(x)$ is of degree 1, that is, $m_j(x) = x - x_j$. From (A.5), we see that, in this case,

$$m_i(x_j)c_{ij}(x_j) = 1. \tag{4.51}$$

Now, any $c_{ij}(x)$ satisfying (4.47) [and hence (4.51)] would do. Choosing the lowest degree, we get

$$c_{ij}(x) = c_{ij} = \frac{1}{m_i(x_j)}. \tag{4.52}$$

With these derivation outlines spelled out, we turn now to specific cases. To establish the evolving pattern, we follow these outlines even in the low-order case ($n = 2$) where direct derivation could be simpler.

4.3.1 Left-Circulant Transformation of Order 2 (Fig. 4.1)

$$\begin{bmatrix} t_1 \\ t_0 \end{bmatrix} = \begin{bmatrix} a_1 & a_0 \\ a_0 & a_1 \end{bmatrix} \begin{bmatrix} b_0 \\ b_1 \end{bmatrix}$$
$$\hat{a}_i = \frac{a_i}{K}; \quad \hat{A}_1(x) = \sum_{i=0}^{1} \hat{a}_i x^i; \quad B_1(x) = \sum_{i=0}^{1} b_i x^i \tag{4.53}$$

Phase 1

$$T_1(x) = K\{\hat{A}_1(x)B_1(x)\} \bmod [\underbrace{(x-1)}_{m_1}\underbrace{(x+1)}_{m_2}]$$ (4.54)

$$u_1(x) = \{\hat{A}_1(x)B_1(x)\} \bmod (x-1)$$

$$= \underbrace{(\hat{a}_0 + \hat{a}_1)}_{\alpha_1}\underbrace{(b_0 + b_1)}_{\beta_1} \quad [\text{see (A.5)}]$$

$$\delta_1 = u_1 = \alpha_1\beta_1$$

$$u_2(x) = \{\hat{A}_1(x)B_1(x)\} \bmod (x+1)$$

$$= \underbrace{(\hat{a}_0 - \hat{a}_1)}_{\alpha_2}\underbrace{(b_0 - b_1)}_{\beta_2}$$

$$\delta_2 = u_2 = \alpha_2\beta_2$$

Phase 2

$$c_{12}(x) = \frac{1}{m_1(x_2)} = \frac{1}{m_1(-1)} = -\frac{1}{2}$$

$$v_1 = \delta_1$$

$$v_2 = \{(\delta_2 - \delta_1)(-\tfrac{1}{2})\} \bmod (x+1) = \tfrac{1}{2}(\delta_1 - \delta_2)$$

$$T_1(x) = K[\delta_1 + \tfrac{1}{2}(\delta_1 - \delta_2)(x-1)] = \underbrace{\left[\frac{K}{2}(\delta_1 - \delta_2)\right]}_{t_1}x + \underbrace{\left[\frac{K}{2}(\delta_1 + \delta_2)\right]}_{t_0}.$$ (4.55)

The obvious choice here is $K = 2$ so that the final result is

$$t_0 = \delta_1 + \delta_2 \; ; \quad t_1 = \delta_1 - \delta_2.$$

The algorithm is summarized graphically in the tableau of Fig. 4.1. The conventions adopted here are quite simple and are also the ones adopted in the more complex tableaus presented later on. We defined β_i as a linear combination of the b_j's. The β_i row lists the (nonzero) coefficients of this linear combination. Similarly, we found that t_i is a linear combination of the δ_j's. The t_i column lists the coefficients of this linear combination. The left corner arrow indicates that the β_i's are derived from the b_j's and not the other way around. Usually there is no directional ambiguity so that the arrows may be omitted. The equations $\delta_i = \beta_i\alpha_i$ appear explicitly in the tableau using the Fortran multiplication symbol.

$$a_i = \phi_i\left(\left\{\frac{a_j}{2}\right\}\right)$$

$$\beta_i = \phi_i\left(\left\{b_j\right\}\right)$$

n = 2 (2M; 4A)   **Fig. 4.1.** Algorithm for LCT of order 2

Finally, recall that the basic equation (4.45) is fully symmetric with respect to $\{\hat{a}_j\}$, $\{b_j\}$. We have taken advantage of this in the terminology introduced here. Thus, coupled with each β_i which is a specific function of $\{b_j\}$ [say, $\phi_i(\{b_j\})$] spelled out in the tableau, there is the variable α_i which is exactly the same function of $\{\hat{a}_j\}$, that is,

$$\alpha_i = \phi_i\left(\left\{\frac{a_j}{K}\right\}\right).$$

This convention is adhered to throughout the chapter. Practically, this means that the left part of the tableau has to be run through twice. First, with $\{a_j/K\}$ replacing $\{b_j\}$ and thus yielding $\{\alpha_i\}$ instead of $\{\beta_i\}$, then we run through the full tableau with $\{b_j\}$ as input. Note, however, that in spite of the mathematical symmetry between $\{a_j/K\}$ and $\{b_j\}$, practically, there is an important difference between them. a is considered a constant matrix transforming a number of different data vectors b. Therefore, the α_i's may be precomputed once and for all, their computation being ignored in accounting for the cost of transforming one data vector b. With this in mind, we count only the number of explicit arithmetic operations in the tableau, arriving at 2 multiplications and 4 additions, for which we adopt the designation (2M; 4A) appearing in the figure. Note that the 2 multiplications are the minimum prescribed by Winograd's theorem (right column in Table 4.1).

4.3.2 Left-Circulant Transformation of Order 4 (Fig. 4.2)

In this and all other tableaus derived in this chapter, it is suggested that the reader consider the tableau as he follows its derivation, noting the graphical representation of each mathematical statement as the algorithm evolves.

$$\begin{bmatrix} t_3 \\ t_2 \\ t_1 \\ t_0 \end{bmatrix} \begin{bmatrix} a_3 & a_2 & a_1 & a_0 \\ a_2 & a_1 & a_0 & a_3 \\ a_1 & a_0 & a_3 & a_2 \\ a_0 & a_3 & a_2 & a_1 \end{bmatrix} \begin{bmatrix} b_0 \\ b_1 \\ b_2 \\ b_3 \end{bmatrix} \qquad \left. \begin{array}{c} \hat{a}_i = \dfrac{a_i}{K} \\[2mm] \hat{A}_3(x) = \displaystyle\sum_{i=0}^{3} \hat{a}_i x^i \\[2mm] B_3(x) = \displaystyle\sum_{i=0}^{3} b_i x^i \end{array} \right\} \qquad (4.56)$$

Phase 1

$$T_3(x) = K\{\hat{A}_3(x)B_3(x)\} \bmod [\underbrace{(x^2+1)}_{m_1}\underbrace{(x+1)}_{m_2}\underbrace{(x-1)}_{m_3}] \tag{4.57}$$

$$u_3(x) = \hat{A}_3(1)B_3(1) = \underbrace{(\hat{a}_0 + \hat{a}_1 + \hat{a}_2 + \hat{a}_3)}_{\alpha_1}\underbrace{(b_0 + b_1 + b_2 + b_3)}_{\beta_1}$$

$$c_1 = b_0 + b_2 ; \quad c_2 = b_1 + b_3$$

$$\therefore \beta_1 = c_1 + c_2$$

$$\delta_1 = u_3 = \alpha_1\beta_1$$

$$u_2(x) = \hat{A}_3(-1)B_3(-1) = \underbrace{(\hat{a}_0 - \hat{a}_1 + \hat{a}_2 - \hat{a}_3)}_{\alpha_2}\underbrace{(b_0 - b_1 + b_2 - b_3)}_{\beta_2}$$

$$\therefore \beta_2 = c_1 - c_2$$

$$\delta_2 = u_2 = \alpha_2\beta_2$$

$u_1(x)$ is evaluated in two steps:

$$\left.\begin{aligned}\hat{A}_3(x)\bmod(x^2+1) &= \underbrace{(\hat{a}_1 - \hat{a}_3)}_{\alpha_3}x + \underbrace{(\hat{a}_0 - \hat{a}_2)}_{\alpha_4} \\ B_3(x)\bmod(x^2+1) &= \underbrace{(b_1 - b_3)}_{\beta_3}x + \underbrace{(b_0 - b_2)}_{\beta_4}\end{aligned}\right\} \quad \text{(from Table A.1).} \tag{4.58}$$

Note that the tableau of Fig. 4.2 defines β_3, β_4 indirectly in terms of

$$c_3 = b_1 - b_3 ; \quad c_4 = b_0 - b_2 .$$

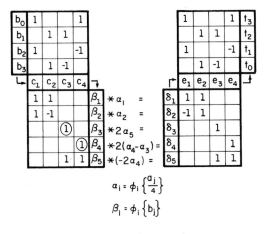

$$\alpha_i = \phi_i\left\{\frac{a_j}{4}\right\}$$

$$\beta_i = \phi_i\{b_j\}$$

$$\underline{n = 4 \ (5M; 15A)}$$

Fig. 4.2. Algorithm for LCT of order 4

Thus,

$$\beta_3 = c_3 ; \quad \beta_4 = c_4$$

so that no arithmetic operations are involved in the last two equations. Whenever this is the case, we circle the relevant terms in the tableau to stress this fact.

Now we combine the two results of (4.58) using Table A.2 ($\alpha_3 = p_1 ; \beta_3 = q_1 ;$ etc.), getting

$$u_1(x) = [-(\alpha_4 - \alpha_3)\beta_4 + \alpha_4(\beta_3 + \beta_4)]x + [\alpha_4(\beta_3 + \beta_4) - (\alpha_3 + \alpha_4)\beta_3].$$

We introduce now

$$\alpha_5 = \alpha_3 + \alpha_4 ; \quad \beta_5 = \beta_3 + \beta_4 = c_3 + c_4$$

$$\delta_3 = 2\alpha_5\beta_3 ; \quad \delta_4 = 2(\alpha_4 - \alpha_3)\beta_4 ; \quad \delta_5 = -2\alpha_4\beta_5 .$$

Hence,

$$u_1(x) = -\tfrac{1}{2}\underbrace{(\delta_4 + \delta_5)}_{e_4}x - \tfrac{1}{2}\underbrace{(\delta_5 + \delta_3)}_{e_3}.$$

Phase 2

$$c_{12} = \frac{1}{m_1(x_2)} = \frac{1}{m_1(-1)} = \frac{1}{2} ; \quad c_{13} = \frac{1}{m_1(x_3)} = \frac{1}{m_1(1)} = \frac{1}{2} ; \quad c_{23} = \frac{1}{m_2(x_3)}$$

$$= \frac{1}{m_2(1)} = \frac{1}{2}$$

$$v_1(x) = u_1(x) = -\tfrac{1}{2}(e_4 x + e_3)$$

$$v_2 = \{(\delta_2 + \tfrac{1}{2}e_4 x + \tfrac{1}{2}e_3)\tfrac{1}{2}\} \bmod (x+1) = \tfrac{1}{4}(2\delta_2 - e_4 + e_3)$$

$$v_3 = \{[(\delta_1 + \tfrac{1}{2}e_4 x + \tfrac{1}{2}e_3)\tfrac{1}{2} - \tfrac{1}{4}(2\delta_2 - e_4 + e_3)]\tfrac{1}{2}\} \bmod (x-1) = \tfrac{1}{4}(\delta_1 - \delta_2 + e_4)$$

$$T_3(x) = -\frac{K}{2}(e_4 x + e_3) + \frac{K}{4}(2\delta_2 - e_4 + e_3)(x^2 + 1) + \frac{K}{4}(\delta_1 - \delta_2 + e_4)$$
$$\cdot (x^3 + x^2 + x + 1).$$

Adopting

$$K = 4,$$

$$T_3(x) = (\delta_1 - \delta_2 + e_4)x^3 + (\delta_1 + \delta_2 + e_3)x^2 + (\delta_1 - \delta_2 - e_4)x + (\delta_1 + \delta_2 - e_3).$$

Introducing now

$$e_1 = \delta_1 - \delta_2 ; \quad e_2 = \delta_1 + \delta_2 ,$$

we get the final result

$$T_3(x) = (e_1 + e_4)x^3 + (e_2 + e_3)x^2 + (e_1 - e_4)x + (e_2 - e_3). \tag{4.59}$$

A count of arithmetic operations in the tableau (excluding the α_i manipulations) yields 5 multiplications and 15 additions[6], again realizing the multiplication minimum of Table 4.1.

4.3.3 Left-Circulant Transformation of Order 6 (Fig. 4.4)

$$
\begin{bmatrix} t_5 \\ t_4 \\ t_3 \\ t_2 \\ t_1 \\ t_0 \end{bmatrix} =
\begin{bmatrix}
a_5 & a_4 & a_3 & a_2 & a_1 & a_0 \\
a_4 & a_3 & a_2 & a_1 & a_0 & a_5 \\
a_3 & a_2 & a_1 & a_0 & a_5 & a_4 \\
a_2 & a_1 & a_0 & a_5 & a_4 & a_3 \\
a_1 & a_0 & a_5 & a_4 & a_3 & a_2 \\
a_0 & a_5 & a_4 & a_3 & a_2 & a_1
\end{bmatrix}
\begin{bmatrix} b_0 \\ b_1 \\ b_2 \\ b_3 \\ b_4 \\ b_5 \end{bmatrix}
\qquad
\begin{aligned}
\hat{a}_i &= \frac{a_i}{K} \\
\hat{A}_5(x) &= \sum_{i=0}^{5} \hat{a}_i x^i \\
B_5(x) &= \sum_{i=0}^{5} b_i x^i.
\end{aligned}
\tag{4.60}
$$

Phase 1

$$T_5(x) = K\{\hat{A}_5(x)B_5(x)\} \bmod \{\underbrace{(x^2 - x + 1)}_{m_1}\underbrace{(x^2 + x + 1)}_{m_2}\underbrace{(x + 1)}_{m_3}\underbrace{(x - 1)}_{m_4}\} \tag{4.61}$$

$$u_4 = \hat{A}_5(1)B_5(1) = \left(\sum_{i=0}^{5} \hat{a}_i\right)\left(\sum_{i=0}^{5} b_i\right)$$

$$c_1 = b_3 + b_0 ; \qquad c_2 = b_5 + b_2 ; \qquad c_3 = b_4 + b_1$$

$$u_4 = \underbrace{\left(\sum_{i=0}^{5} \hat{a}_i\right)}_{\alpha_1}\underbrace{(c_1 + c_2 + c_3)}_{\beta_1}$$

$$\delta_1 = u_4 = \alpha_1\beta_1$$

$$u_3 = \hat{A}_5(-1)B_5(-1) = \underbrace{(-\hat{a}_0 + \hat{a}_1 - \hat{a}_2 + \hat{a}_3 - \hat{a}_4 + \hat{a}_5)}_{\alpha_6}$$

$$\cdot \underbrace{(-b_0 + b_1 - b_2 + b_3 - b_4 + b_5)}_{\beta_6}$$

$$c_6 = b_3 - b_0 ; \qquad c_5 = b_5 - b_2 ; \qquad c_4 = b_4 - b_1$$

$$u_3 = \alpha_6\underbrace{(c_6 + c_5 - c_4)}_{\beta_6}$$

$$\delta_6 = u_3 = \alpha_6\beta_6$$

---

6 The author wishes to acknowledge here the help of Dr. *R. G. Lipes* of JPL in reducing the number of additions from 16 to 15.

$u_1(x), u_2(x)$ are evaluated in two steps

$$B_5(x) \bmod(x^2+x+1) = [\underbrace{(b_4+b_1)}_{c_3} - \underbrace{(b_5+b_2)}_{c_2}]x + [\underbrace{(b_3+b_0)}_{c_1} - \underbrace{(b_5+b_2)}_{c_2}]$$

$$= \underbrace{(c_3-c_2)}_{\beta_2}x + \underbrace{(c_1-c_2)}_{\beta_3} = \beta_2 x + \beta_3 \quad \text{(from Table A.1)}$$

$$\therefore \hat{A}_5(x) \bmod(x^2+x+1) = \alpha_2 x + \alpha_3$$

$$\therefore u_2(x) = \{\hat{A}_5(x)B_5(x)\} \bmod(x^2+x+1)$$

$$= [\alpha_3\beta_3 - (\alpha_3-\alpha_2)(\beta_3-\beta_2)]x + [\alpha_3\beta_3 - \alpha_2\beta_2] \quad (4.62)$$

$$\text{(from Table A.2)}.$$

We introduce now

$$\left.\begin{array}{l} \beta_7 = \beta_3 - \beta_2 = c_1 - c_2 + c_2 - c_3 = c_1 - c_3 \\ \alpha_7 = \alpha_3 - \alpha_2 \end{array}\right\} \tag{4.63}$$

and use these to transform (4.62) into[7]

$$u_2(x) = \underbrace{(\alpha_2\beta_3 + \alpha_7\beta_2)}_{-e_3}x + \underbrace{(\alpha_7\beta_2 + \alpha_3\beta_7)}_{e_2} \tag{4.64}$$

$$B_5(x) \bmod(x^2-x+1) = -[\underbrace{(b_5-b_2)}_{c_5} + \underbrace{(b_4-b_1)}_{c_4}]x - [\underbrace{(b_3-b_0)}_{c_6} - \underbrace{(b_5-b_2)}_{c_5}]$$

$$= -\underbrace{(c_5+c_4)}_{\beta_5}x - \underbrace{(c_6-c_5)}_{\beta_4} = -\beta_5 x - \beta_4$$

$$\text{(from Table A.1)}$$

$$\therefore \hat{A}_5(x) \bmod(x^2-x+1) = -\alpha_5 x - \alpha_4$$

$$\therefore u_1(x) = \{\hat{A}_5(x)B_5(x)\} \bmod(x^2-x+1)$$

$$= [(\alpha_4+\alpha_5)(\beta_4+\beta_5) - \alpha_4\beta_4]x + (\alpha_4\beta_4 - \alpha_5\beta_5) \text{(from Table A.2)}. \tag{4.65}$$

Introducing

$$\left.\begin{array}{l} \beta_8 = \beta_5 + \beta_4 = c_5 + c_4 + c_6 - c_5 = c_4 + c_6 \\ \alpha_8 = \alpha_5 + \alpha_4, \end{array}\right\} \tag{4.66}$$

we transform (4.65) into[7]

$$u_1(x) = \underbrace{(\alpha_5\beta_4 + \alpha_8\beta_5)}_{e_4}x + \underbrace{(\alpha_4\beta_8 - \alpha_8\beta_5)}_{e_5}. \tag{4.67}$$

---

7 The e_i's defined here do not appear in the final tableau and are used only in the intermediate steps.

Phase 2

$$\{m_1(x)c_{12}(x)\} \bmod m_2(x) = 1 .$$

Let $c_{12}(x) = \gamma_1 x + \gamma_0$

$$\therefore m_1(x)c_{12}(x) = (x^2 - x + 1)(\gamma_1 x + \gamma_0) = \gamma_1 x^3 + (\gamma_0 - \gamma_1)x^2 + (\gamma_1 - \gamma_0)x + \gamma_0$$

$$\therefore (\gamma_1 - \gamma_0 - \gamma_0 + \gamma_1)x + (\gamma_0 - \gamma_0 + \gamma_1 + \gamma_1) = 1 \quad \text{(from Table A.1)}$$

$$\therefore \gamma_1 = \gamma_0 = \tfrac{1}{2}; \quad c_{12}(x) = \tfrac{1}{2}(x+1)$$

$$c_{13} = \frac{1}{m_1(x_3)} = \frac{1}{m_1(-1)} = \frac{1}{3}; \quad c_{23} = \frac{1}{m_2(x_3)} = \frac{1}{m_2(-1)} = 1$$

$$c_{14} = \frac{1}{m_1(x_4)} = \frac{1}{m_1(1)} = 1; \quad c_{24} = \frac{1}{m_2(x_4)} = \frac{1}{m_2(1)} = \frac{1}{3}; \quad c_{34} = \frac{1}{m_3(x_4)}$$

$$= \frac{1}{m_3(1)} = \frac{1}{2}$$

$$v_1 = e_4 x + e_5$$

$$v_2 = [(-e_3 x + e_2 - e_4 x - e_5)\tfrac{1}{2}(x+1)] \bmod (x^2 + x + 1)$$

$$v_2 = \tfrac{1}{2}[-(e_4 + e_3)x^2 + (-e_4 - e_3 + e_2 - e_5)x + (e_2 - e_5)] \bmod (x^2 + x + 1)$$

$$v_2 = \tfrac{1}{2}[(e_2 - e_5)x + (e_2 + e_3 + e_4 - e_5)] \quad \text{(from Table A.1)}$$

$$v_3 = \{(\delta_6 - e_4 x - e_5)\tfrac{1}{3} - \tfrac{1}{2}[(e_2 - e_5)x + (e_2 + e_3 + e_4 - e_5)]\} \bmod (x+1)$$

$$= \tfrac{1}{6}(-3e_3 - e_4 - 2e_5 + 2\delta_6)$$

$$v_4 = \tfrac{1}{2}\{\langle(\delta_1 - e_4 x - e_5) - \tfrac{1}{2}[(e_2 - e_5)x + (e_2 + e_3 + e_4 - e_5)]\rangle\tfrac{1}{3}$$
$$- \tfrac{1}{6}(-3e_3 - e_4 - 2e_5 + 2\delta_6)\} \bmod (x-1)$$

$$= \tfrac{1}{6}(\delta_1 - e_2 + e_3 - e_4 + e_5 - \delta_6)$$

$$T_5(x) = K(e_4 x + e_5) + \frac{K}{2}[(e_2 - e_5)x + (e_2 + e_3 + e_4 - e_5)](x^2 - x + 1)$$

$$+ \frac{K}{6}(-3e_3 - e_4 - 2e_5 + 2\delta_6)(x^4 + x^2 + 1)$$

$$+ \frac{K}{6}(\delta_1 - e_2 + e_3 - e_4 + e_5 - \delta_6)(x^5 + x^4 + x^3 + x^2 + x + 1) .$$

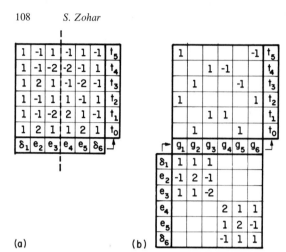

(a)   (b)   Fig. 4.3. Stages in the development of the algorithm for LCT of order 6

Collecting equal power terms yields the desired t_i's. We make now the obvious choice $K=6$ and present the dependence of the t_i's on the e_i's and δ_i's in the tableau format in Fig. 4.3a.

So far, the derivation has been a relatively straightforward application of the guidelines presented at the beginning of the section. However, an overall algorithm based on Fig. 4.3a uses far too many additions. To eliminate some of these, we have to resort to some less obvious manipulations. Examining Fig. 4.3a we note that the coefficients of t_0 and t_3 have identical magnitudes. On the left of the bisecting line, the coefficients themselves are identical, whereas to

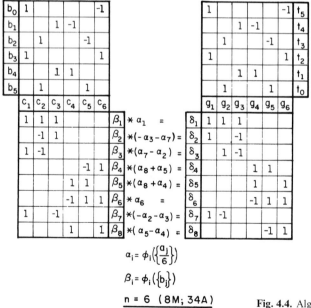

Fig. 4.4. Algorithm for LCT of order 6

the right of it, they have opposite signs. This holds true also for the pairs (t_1, t_4), (t_2, t_5). Taking advantage of these symmetries, we can reduce the number of additions as shown in Fig. 4.3b.

The final manipulations involve the elimination of the e_i's from Fig. 4.3b. This is based on their definitions (4.64, 67) and on a judicious application of (4.63, 66) as follows:

$$g_1-\delta_1=e_3-e_2=-2\alpha_7\beta_2-\alpha_2(\beta_7+\beta_2)-\alpha_3\beta_7=\underbrace{\beta_2(-\alpha_3-\alpha_7)}_{\delta_2}+\underbrace{\beta_7(-\alpha_2-\alpha_3)}_{\delta_7}$$

$$(4.68)$$

$$g_6-\delta_6=e_4-e_5=\alpha_5(\beta_8-\beta_5)+2\alpha_8\beta_5-\alpha_4\beta_8=\underbrace{\beta_5(\alpha_8+\alpha_4)}_{\delta_5}+\underbrace{\beta_8(\alpha_5-\alpha_4)}_{\delta_8} \qquad (4.69)$$

$$g_2-\delta_1=2e_2+e_3=\alpha_7(\beta_3-\beta_7)-\alpha_2\beta_3+2\alpha_3\beta_7=\underbrace{\beta_3(\alpha_7-\alpha_2)}_{\delta_3}-\underbrace{\beta_7(-\alpha_2-\alpha_3)}_{\delta_7} \ (4.70)$$

$$g_5-\delta_6=2e_5+e_4=-\alpha_8(\beta_8-\beta_4)+\alpha_5\beta_4+2\alpha_4\beta_8=\underbrace{\beta_4(\alpha_8+\alpha_5)}_{\delta_4}-\underbrace{\beta_8(\alpha_5-\alpha_4)}_{\delta_8} \ (4.71)$$

$$g_3-\delta_1=-e_2-2e_3=\alpha_7\beta_2+2\alpha_2\beta_3-\alpha_3(\beta_3-\beta_2)=-\underbrace{\beta_2(-\alpha_3-\alpha_7)}_{\delta_2}-\underbrace{\beta_3(\alpha_7-\alpha_2)}_{\delta_3}$$

$$(4.72)$$

$$g_4+\delta_6=e_5+2e_4=\alpha_8\beta_5+2\alpha_5\beta_4+\alpha_4(\beta_5+\beta_4)=\underbrace{\beta_4(\alpha_8+\alpha_5)}_{\delta_4}+\underbrace{\beta_5(\alpha_8+\alpha_4)}_{\delta_5}. \qquad (4.73)$$

This completes the derivation.

4.4 The Basic DFT Algorithms for Prime N

In this section, we apply the tableaus just derived to obtain the DFT algorithms for the odd prime terms of list (4.3), namely, 3, 5, 7. As indicated in Sect. 4.1, these will serve as building blocks for higher order DFT's.

We have seen in Sect. 4.2 that with the proper relabeling, an Nth order DFT matrix displays an nth order LC submatrix (4.12). The main part of the contribution of this submatrix to the overall transformation is spelled out in (4.21) repeated here,

$$B'_\varrho=\Omega\sum_{\sigma=0}^{n-1}W^{(g\varrho+\sigma)}b_\sigma \qquad (\varrho=0,1,...,n-1). \qquad (4.74)$$

On the other hand, the LC tableaus of the last section are based on the (4.29, 39) formulation of the nth order LC transformation. Therefore, in applying the LCT tableaus to the LC transformation expressed in (4.74), we must adopt the following identifications:

$$B'_\varrho = t_{n-1-\varrho} \qquad (\varrho = 0, 1, \ldots, n-1) \tag{4.75}$$

$$a_\varrho = \Omega W^{(g^{n-1-\varrho})} \qquad (\varrho = 0, 1, \ldots, n-1). \tag{4.76}$$

Note the effect of (4.76). The LCT tableaus, being general, provide only a prescription for the computation of the α_i's from the a_i's. However, since (4.76) provides an explicit formula for the a_i's, the α_i's may actually be computed. Specifically, the α_i's are expressible as

$$\alpha_i = \Omega \varepsilon_i \tag{4.77}$$

in which the ε_i's are functions of i, N only and can thus be precomputed once and for all.

We copy now the remaining equations of the relabeled DFT [(4.19, 25, 26)]

$$B_\varrho = \hat{B}_\varrho + B'_\varrho \qquad (\varrho = 0, 1, \ldots, n-1) \tag{4.78}$$

$$\hat{B}_\varrho = \Omega b_{(0)} \qquad (\varrho = 0, 1, \ldots, n-1) \tag{4.79}$$

$$B_{(0)} = \Omega \left(b_{(0)} + \sum_{\sigma=0}^{n-1} b_\sigma \right). \tag{4.80}$$

Note that (4.79, 80) are valid only for the special case of prime N and are based on

$$n = N - 1. \tag{4.81}$$

Equations (4.74–78), on the other hand, are quite general and will also be applied in the next two sections where N is not prime.

Our first step in the construction of the DFT tableau for prime N is the computation of the $\varepsilon_i(N)$ constants. This is done by evaluating the a_ϱ's from (4.76) and then using them in the LCT tableau of order $(N-1)$ to compute the α_i's, and hence the ε_i's (4.77).

The next step is to copy the LCT tableau of order $(N-1)$ into the DFT tableau of order N. This is equivalent to the implementation of (4.74) and yields a tableau transforming the b_i's into the t_j's. Now, using (4.14–16, 75), we replace these variables by f_i's and F_j's. The input and output index sequences we get at this point will usually be nonmonotonic. Therefore, we now permute the rows/columns of the input and output squares to achieve index monotonicity.

It should be pointed out that the variable changes (4.14–16) were introduced to expose the LC structure and thus allow the application of (4.45). Once this has been done, however, we want the resulting DFT tableau to be expressed in terms of the original variables F_u, f_v with standard monotonic index sequences since this simplifies the integration of the tableau into the algorithm for larger N.

We turn now to the implementation of (4.80). Since all three LCT tableaus satisfy

$$\beta_1 = \sum_{i=0}^{n-1} b_i, \qquad (4.82)$$

Equation (4.80) is equivalent to

$$F_0 = B_{(0)} = \Omega[b_{(0)} + \beta_1] = \Omega(f_0 + \beta_1). \qquad (4.83)$$

Finally, in order to efficiently implement (4.78), we examine the three LCT tableaus to see whether they contain a term which appears as a component with coefficient $+1$ of all t_i's. This term turns out to be δ_1.

Hence, replacing $\delta_1 = \alpha_1 \beta_1$ by

$$\delta_1 = \Omega b_{(0)} + \alpha_1 \beta_1 \qquad (4.84)$$

will convert the output from B'_ϱ to B_ϱ. Note, however, that the term $\Omega b_{(0)}$ is already included in $B_{(0)}$ computed in (4.83). This raises the possibility of eliminating one of the two multiplications evident in (4.84). Indeed, combining (4.83) with (4.84) yields [using (4.77)]

$$\delta_1 = B_{(0)} + (\varepsilon_1 - 1)\Omega\beta_1 = F_0 + (\varepsilon_1 - 1)\Omega\beta_1. \qquad (4.85)$$

The multiplier of β_1 is seen here to be the product of Ω and a function of N. This turns out to be general pattern for all β_i in all the DFT tableaus yet to be derived. Hence, we introduce now the notation

$$\xi_i = (\omega_i \Omega)\beta_i, \qquad (4.86)$$

where ω_i is a function of i, N only. β_i is always initially transformed as in (4.86). Hence one of our tasks in constructing the DFT tableaus is the determination of the numerical values of the ω_i constants. As we shall see in the course of the developments, the underlying LCT tableaus in conjunction with (4.77, 86) always uniquely determine the ω_i's. In the case of (4.85)

$$\omega_1 = \varepsilon_1 - 1 \; ; \quad \delta_1 = B_{(0)} + (\omega_1 \Omega)\beta_1 = F_0 + (\omega_1 \Omega)\beta_1 \, . \tag{4.87}$$

So far, we have considered the LCT-DFT transition in general terms. We turn now to specific cases.

4.4.1 DFT of Order 3 (Fig. 4.5)

Equation (4.83) is explicitly stated in the tableau. From (4.76) $\left[\text{with } W = \exp\left(-i\dfrac{2\pi}{3}\right); g = 2\right]$ we get

$$\begin{bmatrix} \dfrac{a_0}{2} \\[2mm] \dfrac{a_1}{2} \end{bmatrix} = \dfrac{\Omega}{2} \begin{bmatrix} W^2 \\[1mm] W^1 \end{bmatrix} = \dfrac{\Omega}{2} \begin{bmatrix} \bar{W}^1 \\[1mm] W^1 \end{bmatrix} , \tag{4.88}$$

where $\bar{W}$ is the complex conjugate of W. Hence, applying the second-order LCT tableau (Fig. 4.1), we find

$\dfrac{\Omega}{2}\bar{W}^1$	$\dfrac{\Omega}{2}W^1$		
1	1	α_1	$= -\dfrac{\Omega}{2}\; ; \quad \varepsilon_1 = -\dfrac{1}{2}\; ; \quad \omega_1 = \varepsilon_1 - 1 = -\dfrac{3}{2}$
1	-1	α_2	$= i\dfrac{\sqrt{3}}{2}\Omega\; ; \quad \varepsilon_2 = i\dfrac{\sqrt{3}}{2}\; ; \quad \omega_2 = \varepsilon_2 = i\dfrac{\sqrt{3}}{2}$

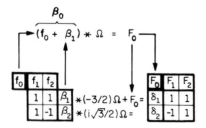

$$N = 3 \; (3M \, ; 6A)$$

Fig. 4.5. Algorithm for DFT of order 3

4.4.2 DFT of Order 5 (Fig. 4.6)

$W = \exp\left(-i\dfrac{2\pi}{5}\right)$; $g = 2$. Hence, from (4.76),

$$\frac{1}{4}\begin{bmatrix} a_0 \\ a_1 \\ a_2 \\ a_3 \end{bmatrix} = \frac{\Omega}{4}\begin{bmatrix} W^3 \\ W^4 \\ W^2 \\ W^1 \end{bmatrix} = \frac{\Omega}{4}\begin{bmatrix} \bar{W}^2 \\ \bar{W}^1 \\ W^2 \\ W^1 \end{bmatrix}. \tag{4.89}$$

The α_i's are determined from the following prescription of Fig. 4.2:

$\dfrac{\Omega}{4}\bar{W}^2$	1			1	$c_1 = -\dfrac{\Omega}{2}\cos 36°$
$\dfrac{\Omega}{4}\bar{W}^1$		1	1		$c_2 = \dfrac{\Omega}{2}\cos 72°$
$\dfrac{\Omega}{4}W^2$	1			-1	$c_3 = \dfrac{i\Omega}{2}\sin 72°$
$\dfrac{\Omega}{4}W^1$		1	-1		$c_4 = \dfrac{i\Omega}{2}\sin 36°$
	c_1	c_2	c_3	c_4	

1	1			α_1 $= -\dfrac{\Omega}{4}$
1	-1			α_2 $= -\dfrac{\Omega}{2}(\cos 36° + \cos 72°)$
		①		α_3 $= \dfrac{i\Omega}{2}\sin 72°$
			①	α_4 $= \dfrac{i\Omega}{2}\sin 36°$
		1	1	α_5 $= \dfrac{i\Omega}{2}(\sin 36° + \sin 72°)$

(4.90)

Equations (4.77, 90) prescribe the ε_i's. Finally, from (4.87) and the β_i multipliers in Fig. 4.2, we get

$$
\left.
\begin{aligned}
\omega_1 &= \varepsilon_1 - 1 = -\tfrac{5}{4} \\
\omega_2 &= \varepsilon_2 = -\tfrac{1}{2}(\cos 36° + \cos 72°) \\
\omega_3 &= 2\varepsilon_5 = i(\sin 36° + \sin 72°) \\
\omega_4 &= 2(\varepsilon_4 - \varepsilon_3) = i(\sin 36° - \sin 72°) \\
\omega_5 &= -2\varepsilon_4 = -i \sin 36° .
\end{aligned}
\right\}
\tag{4.91}
$$

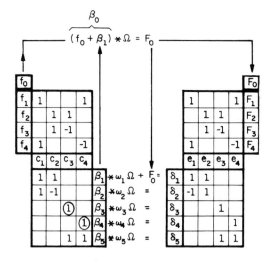

$$N = 5 \quad (6M ; 17A)$$

Fig. 4.6. Algorithm for DFT of order 5

4.4.3 DFT of Order 7 (Fig. 4.8)

$$W = \exp\left(-i\,\frac{2\pi}{7}\right); \; g = 3. \text{ Hence from (4.76)}$$

$$
\frac{1}{6}
\begin{bmatrix}
a_0 \\ a_1 \\ a_2 \\ a_3 \\ a_4 \\ a_5
\end{bmatrix}
=
\frac{\Omega}{6}
\begin{bmatrix}
W^5 \\ W^4 \\ W^6 \\ W^2 \\ W^3 \\ W^1
\end{bmatrix}
=
\frac{\Omega}{6}
\begin{bmatrix}
\bar{W}^2 \\ \bar{W}^3 \\ \bar{W}^1 \\ W^2 \\ W^3 \\ W^1
\end{bmatrix}.
\tag{4.92}
$$

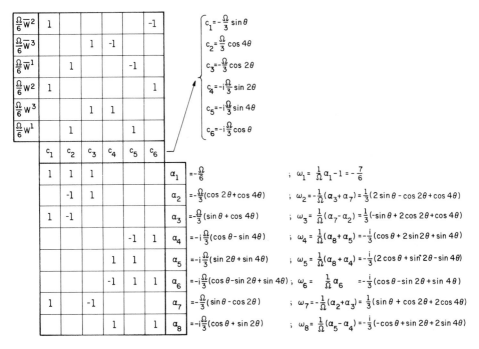

Fig. 4.7. Computation of ω_i's for the DFT algorithm of order 7

All α_i's are expressible in terms of the angle

$$\theta = \frac{90°}{7} \tag{4.93}$$

and its multiples. The prescription of Fig. 4.4 for the computation of the α_i's is shown in Fig. 4.7 and the final tableau based on that is shown in Fig. 4.8.

4.5 The Basic DFT Algorithms for $N=4, 9$

From (4.2) we see that with the tableaus of the last section, the highest realizable N would be $105(=3 \cdot 5 \cdot 7)$. With this in mind, we add now the trivial algorithm for $N=2$ (Fig. 4.9), thus increasing the maximal N to 210.

If we want still higher N, we may either generate new tableaus for successively higher primes $(11, 13, 17, \ldots)$, or devise new tableaus for $N=p^k$ (p prime). We adopt the latter course here.

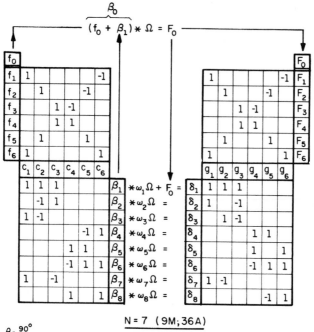

$$N = 7 \quad (9M; 36A)$$

$\theta = \dfrac{90°}{7}$

$\omega_1 = -\dfrac{7}{6}$; $\omega_5 = -\dfrac{1}{3}(2\cos\theta + \sin 2\theta - \sin 4\theta)$

$\omega_2 = \dfrac{1}{3}(2\sin\theta - \cos 2\theta + \cos 4\theta)$; $\omega_6 = -\dfrac{1}{3}(\cos\theta - \sin 2\theta + \sin 4\theta)$

$\omega_3 = \dfrac{1}{3}(-\sin\theta + 2\cos 2\theta + \cos 4\theta)$; $\omega_7 = \dfrac{1}{3}(\sin\theta + \cos 2\theta + 2\cos 4\theta)$

$\omega_4 = -\dfrac{1}{3}(\cos\theta + 2\sin 2\theta + \sin 4\theta)$; $\omega_8 = -\dfrac{1}{3}(-\cos\theta + \sin 2\theta + 2\sin 4\theta)$

Fig. 4.8. Algorithm for DFT of order 7

	f_0	f_1
	1	1	β_0 $*\Omega = $ F_0
	1	-1	β_1 $*\Omega = $ F_1

$$N = 2 \quad (2M; 2A)$$

Fig. 4.9. Algorithm for DFT of order 2

4.5.1 DFT of Order 4 (Fig. 4.10)

$N = 2^2$, yielding (4.12),

$$n = 2. \tag{4.94}$$

The primitive root of 4 is 3. Hence,

ϱ	0	1
$r = 3^{\varrho} \bmod 4$	1	3

$$\tag{4.95}$$

The number of rows excluded from the LC pattern is 2. This number increases still further in the subsequent tableaus, reaching 8 for $N = 16$. This calls for some new and some modified terminology to facilitate handling the non-LC part of the matrix.

First we extend the definition of r so that (4.15) will now include (4.16)

ϱ	0	1	(0)	(2)
r	1	3	0	2

$$\tag{4.96}$$

Similarly, for the column indices σ, s, we now have

σ	0	1	(0)	(2)
s	1	3	0	2

$$\tag{4.97}$$

In subsequent derivations, here and in the next section, we shall assume without explicitly so stating that the definitions of r, s have been extended, as indicated here, to cover all indices. Next, we complement (4.19–21) with

$$F'_r = B'_\varrho ; \quad \hat{F}_r = \hat{B}_\varrho ; \quad F_r = \hat{F}_r + F'_r . \tag{4.98}$$

Finally, recalling that the (r, s) element of the DFT matrix (4.9) is ΩW^{rs}, we introduce the "exponent matrix" E,

$$E_{r,s} = (rs) \bmod N \tag{4.99}$$

which we find very convenient in accounting for the contribution of the non-LC part.

In the present case

$$
\begin{array}{llll}
s\to & 0 & 2 & 1 & 3 \\
\sigma\to & (0) & (2) & 0 & 1
\end{array}
$$

$$
E=\left[\begin{array}{cc|cc}
0 & 0 & 0 & 0 \\
0 & 0 & 2 & 2 \\
\hline
0 & 2 & 1 & 3 \\
0 & 2 & 3 & 1
\end{array}\right]
\begin{array}{cc}
(0) & 0 \\
(2) & 2 \\
0 & 1 \\
1 & 3 \\
\uparrow & \uparrow \\
\varrho & r
\end{array}
\tag{4.100}
$$

Note that we have added here both kinds of row and column indices. One simple way of obtaining E is directly from its definition (4.99). In writing it down, we make sure that the nonbracketed ϱ, σ indices would follow the sequence $0, 1, 2, \dots$. This will bring out the LC structure. The sequence of the other indices is immaterial.

We observe that (4.100) displays a second-order LC submatrix as was to be expected. We handle the effect of this submatrix with the LCT tableau of Fig. 4.1, starting with the evaluation of the α_i's

$$
\frac{1}{2}\begin{bmatrix} a_0 \\ a_1 \end{bmatrix} = \frac{\Omega}{2}\begin{bmatrix} W^3 \\ W^1 \end{bmatrix} = \frac{\Omega}{2}\begin{bmatrix} i \\ -i \end{bmatrix}.
\tag{4.101}
$$

Hence (from Fig. 4.1)

$$
\alpha_1 = 0 ; \qquad \alpha_2 = i\Omega .
\tag{4.102}
$$

We conclude that only the β_2 row contributes to F_i' and we copy it into the β_3 row of Fig. 4.10.

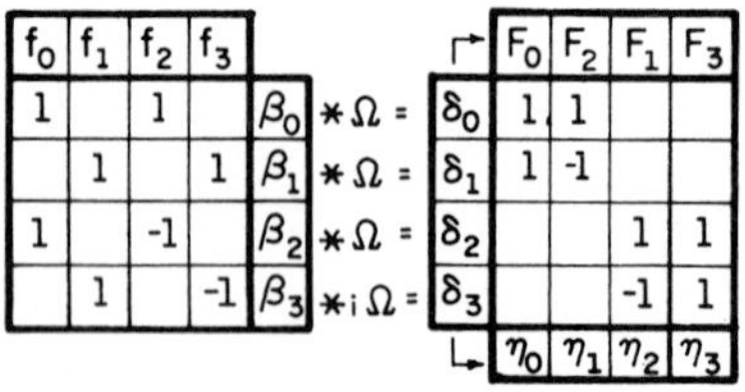

N = 4 (4M; 8A)

Fig. 4.10. Algorithm for DFT of order 4

Note that the output vector F appears scrambled in Fig. 4.10. In Sect. 4.9 we point out the advantage of certain tableau structures which allow storage economies in implementing the algorithm. The tableaus generated up to this point have both this desirable structure and an unscrambled output F. From here on, it seems impossible to realize both these desirable features simultaneously and we opt for the more important memory-conserving structure. Hence, the scrambled output. The scrambling here is quite simple but becomes complex for $N=16$. To facilitate prescribing and handling of this, we have added the vector η to the affected tableaus. The scrambled F is identical with the unscrambled η (see Fig. 4.10).

We return now to the realization of the remainder of the E matrix (4.100)

$$\left. \begin{aligned} F_r - F'_r &= \underbrace{\Omega(f_0 - f_2)}_{\beta_2} = \underbrace{\Omega\beta_2}_{\delta_2} \\ F_r &= F'_r + \delta_2 \end{aligned} \right\} \quad (r=1,3). \tag{4.103}$$

We have used here the fact that for $N=4$, $W^2 = -1$. Equations like (4.103) can be read off directly from the E matrix. Such equations will henceforth be presented without any comment.

$$F_2 = \Omega[\underbrace{(f_0+f_2)}_{\beta_0} - \underbrace{(f_1+f_3)}_{\beta_1}] = \underbrace{\Omega\beta_0}_{\delta_0} - \underbrace{\Omega\beta_1}_{\delta_1} = \delta_0 - \delta_1$$

$$F_0 = \Omega[\underbrace{(f_0+f_2)}_{\beta_0} + \underbrace{(f_1+f_3)}_{\beta_1}] = \underbrace{\Omega\beta_0}_{\delta_0} + \underbrace{\Omega\beta_1}_{\delta_1} = \delta_0 + \delta_1 .$$

This completes the derivation.

4.5.2 DFT of Order 9 (Fig. 4.13)

$N=3^2$. Hence (4.12),

$$n=6. \tag{4.104}$$

The primitive root of 9 is the primitive root of 3, namely, 2. This leads to

ϱ	0	1	2	3	4	5
$r=2^\varrho \bmod 9$	1	2	4	8	7	5

$$\tag{4.105}$$

Hence, the following E matrix:

$$
\begin{array}{ccc}
s\rightarrow\ \ 0 \quad 3 \quad 6 \quad 1 \quad 2 \quad 4 \quad 8 \quad 7 \quad 5 \\
\sigma\rightarrow\ (0)\ (3)\ (6)\ \ 0 \quad 1 \quad 2 \quad 3 \quad 4 \quad 5
\end{array}
$$

$$
E = \left[\begin{array}{ccc|cccccc}
0 & 0 & 0 & 0 & 0 & 0 & 0 & 0 & 0 \\
0 & 0 & 0 & 3 & 6 & 3 & 6 & 3 & 6 \\
0 & 0 & 0 & 6 & 3 & 6 & 3 & 6 & 3 \\
\hline
0 & 3 & 6 & 1 & 2 & 4 & 8 & 7 & 5 \\
0 & 6 & 3 & 2 & 4 & 8 & 7 & 5 & 1 \\
0 & 3 & 6 & 4 & 8 & 7 & 5 & 1 & 2 \\
0 & 6 & 3 & 8 & 7 & 5 & 1 & 2 & 4 \\
0 & 3 & 6 & 7 & 5 & 1 & 2 & 4 & 8 \\
0 & 6 & 3 & 5 & 1 & 2 & 4 & 8 & 7
\end{array}\right]
\begin{array}{cc}
(0) & 0 \\
(3) & 3 \\
(6) & 6 \\
0 & 1 \\
1 & 2 \\
2 & 4 \\
3 & 8 \\
4 & 7 \\
5 & 5 \\
\uparrow & \uparrow \\
\varrho & r
\end{array}
\qquad (4.106)
$$

With $n=6$, Fig. 4.4 will be applicable here. We consider the α_i's first. The E matrix (4.106) clearly displays the a_i sequence [this is also in agreement with (4.76)]. Hence,

$$
\frac{1}{6}\begin{bmatrix} a_0 \\ a_1 \\ a_2 \\ a_3 \\ a_4 \\ a_5 \end{bmatrix} = \frac{\Omega}{6}\begin{bmatrix} W^5 \\ W^7 \\ W^8 \\ W^4 \\ W^2 \\ W^1 \end{bmatrix} = \frac{\Omega}{6}\begin{bmatrix} \bar{W}^4 \\ \bar{W}^2 \\ \bar{W}^1 \\ W^4 \\ W^2 \\ W^1 \end{bmatrix}. \qquad (4.107)
$$

The computation of the α_i's and ω_i's is shown in Fig. 4.11. Note that, in the terminology of Figs. 4.4, 11,

$$
\alpha_1 = \alpha_6 = 0. \qquad (4.108)
$$

This means that in the realization of F_i' which is effected by copying Figs. 4.4, 11 into Fig. 4.13, the terms associated with α_1, α_6, should be eliminated[8]. In the actual copying from these figures, we also apply some permutations in addition

---

8 Figure 4.13 seems to indicate that these terms have not been eliminated. This is misleading as the terms supporting this impression are those added later on.

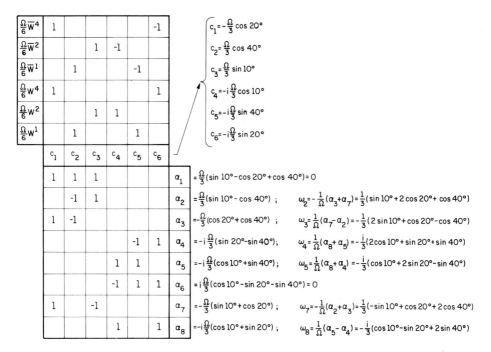

Fig. 4.11. Computation of ω_i's for the DFT algorithm of order 9

Figs. 4.4, 11 index (i)	0	1	2	3	4	5	6	7	8	
$b_i \to f_j$	1	2	4	8	7	5				
$c_i \to c_j$		1	4	2	7	5	8			
$\omega_i \to \omega_j;\ \beta_i \to \beta_j;\ \delta_i \to \delta_j$			4	2	7	5	9	10		*Fig. 4.13* index (j)
$g_i \to e_j$		9	4	2	7	5	10			
$e_j \to g_j$		3	4	2	7	5	6			
$t_i \to F_j$	5	7	8	4	2	1				

Fig. 4.12. Index permutations in assembling Fig. 4.13

to the obvious ones involving the input and output variables. These permutations are spelled out in Fig. 4.12. Note that the permutation $g_i \to g_j$ is shown in two sequential steps: $g_i \to e_j$; $e_j \to g_j$. (Example: $g_1 \to e_9 \to g_3$)[9]

---

9 In the case of $g_3 \to e_2 \to g_2$, we introduce two canceling sign changes.

We turn now to account for the rest of the matrix using (4.106) as our guide. In doing that we encounter the following constants:

$$\Omega W^3 = \Omega(-\sin 30° - i\cos 30°) = -\Omega\left(\frac{1}{2} + i\frac{\sqrt{3}}{2}\right)$$

$$\Omega W^6 = \Omega\bar{W}^3 = -\Omega\left(\frac{1}{2} - i\frac{\sqrt{3}}{2}\right)$$

which we use in the following.

$r=1;\ 4;\ 7$

$$F_r - F_r' = \Omega\left[f_0 - \left(\frac{1}{2} + i\frac{\sqrt{3}}{2}\right)f_3 - \left(\frac{1}{2} - i\frac{\sqrt{3}}{2}\right)f_6\right]$$

$$= \Omega\left[\underbrace{f_0}_{\beta_0} - \underbrace{\tfrac{1}{2}(f_6+f_3)}_{\beta_3} + i\frac{\sqrt{3}}{2}\underbrace{(f_6-f_3)}_{\beta_6}\right] = \underbrace{\Omega\beta_0}_{\delta_0} - \underbrace{\tfrac{1}{2}\Omega\beta_3}_{\delta_3} - \underbrace{\left(-i\frac{\sqrt{3}}{2}\Omega\beta_6\right)}_{\delta_6}$$

$$= \underbrace{\delta_0 - \delta_3}_{e_3} - \underbrace{\delta_6}_{e_6}$$

$$\therefore\ F_r = F_r' + e_3 - e_6.$$

We withhold implementation and proceed to the next group

$r=2;\ 8;\ 5.$

Compared to the previous case we have here interchange of W^3 with $W^6(=\bar{W}^3)$. Hence,

$$F_r = F_r' + e_3 + e_6.$$

At this point we implement both groups as shown in Fig. 4.13.

Next we implement F_3, F_6. From (4.106),

$$F_3 = \Omega\left[(f_0+f_3+f_6) - \left(\frac{1}{2}+i\frac{\sqrt{3}}{2}\right)(f_1+f_4+f_7) - \left(\frac{1}{2}-i\frac{\sqrt{3}}{2}\right)(f_2+f_8+f_5)\right]$$

$$= \Omega\left\{\underbrace{f_0}_{\beta_0} + \underbrace{(f_6+f_3)}_{\beta_3} - \tfrac{1}{2}[\underbrace{(f_8+f_1)}_{c_1} + \underbrace{(f_7+f_2)}_{c_2} + \underbrace{(f_5+f_4)}_{c_4}]\right.$$

$$\left. + i\frac{\sqrt{3}}{2}[\underbrace{(f_8-f_1)}_{c_8} - \underbrace{(f_7-f_2)}_{c_7} + \underbrace{(f_5-f_4)}_{c_5}]\right\}$$

$$= \underbrace{\Omega\beta_0}_{\delta_0} + \underbrace{\Omega\beta_3}_{2\delta_3} - \tfrac{1}{2}\Omega\underbrace{(c_1+c_2+c_4)}_{\beta_1} + i\frac{\sqrt{3}}{2}\Omega\underbrace{(c_5-c_7+c_8)}_{\beta_8}$$

$$= \underbrace{(\delta_0+2\delta_3)}_{e_0} - \underbrace{(\tfrac{1}{2}\Omega\beta_1)}_{\delta_1} - \underbrace{\left(-i\frac{\sqrt{3}}{2}\Omega\beta_8\right)}_{\delta_8}$$

$$= e_0 - \underbrace{\delta_1}_{e_1} - \underbrace{\delta_8}_{e_8} = \underbrace{e_0-e_1}_{g_1} - \underbrace{e_8}_{g_8} = g_1 - g_8.$$

	c_0	c_1	c_2	c_3	c_4	c_5	c_6	c_7	c_8
f_0	(1)								
f_1		1							-1
f_2			1					-1	
f_3				1			-1		
f_4					1	-1			
f_5					1	1			
f_6				1			1		
f_7			1					1	
f_8		1							1

c_0	c_1	c_2	c_3	c_4	c_5	c_6	c_7	c_8			
(1)									β_0	$*\Omega$	$= \delta_0$
	1	1		1					β_1	$*(1/2)\Omega$	$= \delta_1$
	1			-1					β_2	$*\omega_2\Omega$	$= \delta_2$
			(1)						β_3	$*(1/2)\Omega$	$= \delta_3$
		1		-1					β_4	$*\omega_4\Omega$	$= \delta_4$
					1		1		β_5	$*\omega_5\Omega$	$= \delta_5$
						(1)			β_6	$*(-i\sqrt{3}/2)\Omega$	$= \delta_6$
					-1			1	β_7	$*\omega_7\Omega$	$= \delta_7$
					1		-1	1	β_8	$*(-i\sqrt{3}/2)\Omega$	$= \delta_8$
	1	-1							β_9	$*\omega_9\Omega$	$= \delta_9$
							1	1	β_{10}	$*\omega_{10}\Omega$	$= \delta_{10}$

	e_0	e_1	e_2	e_3	e_4	e_5	e_6	e_7	e_8	e_9	e_{10}
δ_0	1			1							
δ_1		(1)									
δ_2			1		1						
δ_3	2			-1							
δ_4			1							1	
δ_5								1			1
δ_6							(1)				
δ_7						1		1			
δ_8									(1)		
δ_9					-1					1	
δ_{10}						-1					1

	e_0	e_1	e_2	e_3	e_4	e_5	e_6	e_7	e_8	e_9	e_{10}
g_0	1	2									
g_1	1	-1									
g_2			-1	1							
g_3				1						1	
g_4				1	1						
g_5						1	1				
g_6							1				1
g_7							-1	1			
g_8									(1)		

	F_0	F_3	F_2	F_1	F_4	F_5	F_8	F_7	F_6
g_0	(1)								
g_1		1							1
g_2			1					1	
g_3				1			1		
g_4					1	1			
g_5					-1	1			
g_6				-1			1		
g_7			-1					1	
g_8		-1							1
	η_0	η_1	η_2	η_3	η_4	η_5	η_6	η_7	η_8

$\omega_2 = -\frac{1}{3}(2\sin 10° + \cos 20° - \cos 40°)$; $\omega_7 = -\frac{i}{3}(2\cos 10° + \sin 20° + \sin 40°)$

$\omega_4 = \frac{1}{3}(\sin 10° + 2\cos 20° + \cos 40°)$; $\omega_9 = \frac{1}{3}(-\sin 10° + \cos 20° + 2\cos 40°)$

$\omega_5 = -\frac{i}{3}(\cos 10° + 2\sin 20° - \sin 40°)$; $\omega_{10} = -\frac{i}{3}(\cos 10° - \sin 20° + 2\sin 40°)$

N = 9 (11M; 44A)

Fig. 4.13. Algorithm for DFT of order 9

We turn now to F_6. The only difference here is the interchange of W^3 with $W^6 (= \bar{W}^3)$. Hence

$$F_6 = g_1 + g_8 .$$

Finally,

$$F_0 = \Omega(f_0 + f_3 + f_6) + \Omega\beta_1 = \underbrace{\Omega\beta_0}_{\delta_0} + \underbrace{\Omega\beta_3}_{2\delta_3} + \underbrace{\Omega\beta_1}_{2\delta_1} = \underbrace{(\delta_0 + 2\delta_3)}_{e_0} + \underbrace{2\delta_1}_{e_1}$$
$$= e_0 + 2e_1 = g_0 .$$

This completes the derivation.

With the two additional tableaus developed in this section, the maximal realizable N has been pushed to $1260 (4 \cdot 5 \cdot 7 \cdot 9)$. We push it still higher (5040) with the tableaus of the next section.

4.6 The Basic DFT Algorithms for $N = 8, 16$

In developing the tableaus for $N = 8, 16$ we face a complication due to the fact that

$$N = 2^k \quad (k > 2) \tag{4.109}$$

has no primitive roots. In other words, there is no integer whose powers $(\mod N)$ would generate all of the odd numbers in the interval $(1, N)$. We can, however, generate one half of these with powers $(\mod N)$ of the number 3 and the other half with the negatives of these powers $(\mod N)$ [4.9]. Thus, we are led to a partitioning of $\{F_r\}$ into 3 subsets,

$$F_r = \begin{cases} B_{(r)} & (r \text{ even}) \\ B_\varrho & (r = 3^\varrho \mod N) \\ B_{\bar{\varrho}} & (r = -3^\varrho \mod N). \end{cases} \tag{4.110}$$

As in the case of the parenthesized index (i) introduced in Sect. 4.2, the barred index $\bar{i}$ serves to identify the subset to which $B_{\bar{i}}$ belongs. Note that the index zero appears in all three subsets. Thus,

$$B_{(0)} = F_0 ; \quad B_0 = F_1 ; \quad B_{\bar{0}} = F_{N-1} . \tag{4.111}$$

We conclude that (4.14) should now be replaced by

$$\left. \begin{array}{l} r = 3^\varrho \mod N \\ s = 3^\sigma \mod N \end{array} \right\} \quad \left(\varrho, \sigma = 0, 1, \dots \frac{N}{4} - 1 \right) \tag{4.112}$$

$$\left. \begin{array}{l} r = -3^\varrho \mod N \\ s = -3^\sigma \mod N \end{array} \right\} \quad \left(\varrho, \sigma = \bar{0}, \bar{1}, \dots \frac{\overline{N}}{4} - 1 \right). \tag{4.113}$$

We illustrate this with the case $N=8$

ϱ	$\bar{0}$	$\bar{1}$	0	1
$r=3^{\varrho} \bmod 8$	–	–	1	3
$r=-3^{\varrho} \bmod 8$	7	5	–	–

(4.114)

Equations (4.15, 16) should now read

$$\left. \begin{array}{l} F_r=B_{\varrho}; f_s=b_{\sigma} \ (r, s \text{ odd}) \\ F_r=B_{(r)}; f_s=b_{(s)} \ (r, s \text{ even}) \end{array} \right\} \tag{4.115}$$

so that the overall r, ϱ functional relationship for $N=8$ can be summarized as follows:

ϱ	$\bar{0}$	$\bar{1}$	0	1	(0)	(2)	(4)	(6)
r	7	5	1	3	0	2	4	6

(4.116)

with an identical table relating s to σ.

Equations (4.112–115) are now applied to eliminate F_u, f_v from (4.9). (See analogous treatment in Sect. 4.2 and note $p=2$ here.)

$$B_{(2t)}=F_{2t}=\Omega \sum_{m=0}^{\frac{N}{2}-1} W^{4mt} b_{(2m)} + \Omega \sum_{\sigma=0}^{\frac{N}{4}-1} (W^{-2t3^{\sigma}} b_{\bar{\sigma}} + W^{2t3^{\sigma}} b_{\sigma}) \ \left(t=0,1,\dots\frac{N}{2}-1\right) \tag{4.117}$$

$$B_{\varrho}=\hat{B}_{\varrho}+B'_{\varrho}; \quad \hat{F}_r=\hat{B}_{\varrho}; \quad F'_r=B'_{\varrho} \quad (r \text{ odd}) \tag{4.118}$$

$$\left. \begin{array}{l} \hat{B}_{\bar{\varrho}}=\Omega \displaystyle\sum_{m=0}^{\frac{N}{2}-1} W^{-2m3^{\varrho}} b_{(2m)} \\[2em] \hat{B}_{\varrho}=\Omega \displaystyle\sum_{m=0}^{\frac{N}{2}-1} W^{2m3^{\varrho}} b_{(2m)} \end{array} \right\} \quad \left(\varrho=0,1,\dots\frac{N}{4}-1\right) \tag{4.119}$$

$$\left. \begin{array}{l} B'_{\bar{\varrho}}=\Omega \displaystyle\sum_{\sigma=0}^{\frac{N}{4}-1} [W^{(3\varrho+\sigma)} b_{\bar{\sigma}} + W^{-(3\varrho+\sigma)} b_{\sigma}] \\[2em] B'_{\varrho}=\Omega \displaystyle\sum_{\sigma=0}^{\frac{N}{4}-1} [W^{-(3\varrho+\sigma)} b_{\bar{\sigma}} + W^{(3\varrho+\sigma)} b_{\sigma} \end{array} \right\} \quad \left(\varrho=0,1,\dots,\frac{N}{4}-1\right). \tag{4.120}$$

It is obvious that all four matrix products appearing in (4.120) are Hankel transformations $(\varrho + \sigma)$ of order $N/4$. We prove now that they are also LCT's. To do this, we must show that [see (4.6, 22)]

$$3^{\varrho + \frac{N}{4} - 1} = 3^{\varrho - 1} \bmod N \qquad (N = 2^k; \, k > 2) \tag{4.121}$$

or equivalently,

$$3^{\frac{N}{4}} = 1 \bmod N \qquad (N = 2^k; \, k > 2). \tag{4.122}$$

We prove (4.122) by induction on k. Assume it to be true for $N = n$, then $3^{n/4} = mn + 1$ (m integer). Squaring yields $3^{n/2} = \left(m^2 \frac{n}{2}\right)(2n) + m(2n) + 1$ so that $3^{(2n)/4} = 1 \bmod (2n)$ and (4.122) is true for $N = 2n$. Finally, (4.122) is obviously true for $k = 3$.

Thus, (4.120) involves four LC matrix products. Furthermore, these four LC matrices comprise a second-order compound matrix which is an LC itself. All these features stand out quite clearly in the two cases to be developed now.

4.6.1 DFT of Order 8 (Fig. 4.14)

Applying the permutation of (4.116), we get the following E matrix:

$$
\begin{array}{cccccccc}
s \to & 0 & 4 & 2 & 6 & 7 & 5 & 1 & 3 \\
\sigma \to & (0) & (4) & (2) & (6) & \bar{0} & \bar{1} & 0 & 1
\end{array}
$$

$$
E = \left[
\begin{array}{cccc|cccc}
0 & 0 & 0 & 0 & 0 & 0 & 0 & 0 \\
0 & 0 & 0 & 0 & 4 & 4 & 4 & 4 \\
0 & 0 & 4 & 4 & 6 & 2 & 2 & 6 \\
0 & 0 & 4 & 4 & 2 & 6 & 6 & 2 \\
\hline
0 & 4 & 6 & 2 & 1 & 3 & 7 & 5 \\
0 & 4 & 2 & 6 & 3 & 1 & 5 & 7 \\
0 & 4 & 2 & 6 & 7 & 5 & 1 & 3 \\
0 & 4 & 6 & 2 & 5 & 7 & 3 & 1
\end{array}
\right]
\begin{array}{cc}
(0) & 0 \\
(4) & 4 \\
(2) & 2 \\
(6) & 6 \\
\bar{0} & 7 \\
\bar{1} & 5 \\
0 & 1 \\
1 & 3 \\
\uparrow & \uparrow \\
\varrho & r
\end{array} \tag{4.123}
$$

We turn now to an explicit representation of the submatrix corresponding to the lower right quarter of E. Denoting

$$w_k = \Omega W^k, \tag{4.124}$$

we get

$$
\begin{aligned}
\hat{t}_1 \left\{ \begin{bmatrix} F'_7 \\ F'_5 \end{bmatrix} \right. \\
\hat{t}_0 \left\{ \begin{bmatrix} F'_1 \\ F'_3 \end{bmatrix} \right.
\end{aligned}
\overset{\overbrace{}^{\hat{a}_1}\;\overbrace{}^{\hat{a}_0}}{
\begin{bmatrix}
w_1 & w_3 & w_7 & w_5 \\
w_3 & w_1 & w_5 & w_7 \\
w_7 & w_5 & w_1 & w_3 \\
w_5 & w_7 & w_3 & w_1
\end{bmatrix}}
\begin{bmatrix} f_7 \\ f_5 \\ f_1 \\ f_3 \end{bmatrix}
\begin{aligned} \left. \right\} \hat{b}_0 \\ \left. \right\} \hat{b}_1 \end{aligned}
\tag{4.125}
$$

We see here four second-order LC submatrices, two each, of types $\hat{a}_0$, $\hat{a}_1$. Using the terminology introduced here, we write down the equivalent second-order block matrix

$$
\begin{bmatrix} \hat{t}_1 \\ \hat{t}_0 \end{bmatrix} = \begin{bmatrix} \hat{a}_1 & \hat{a}_0 \\ \hat{a}_0 & \hat{a}_1 \end{bmatrix} \begin{bmatrix} \hat{b}_0 \\ \hat{b}_1 \end{bmatrix}.
\tag{4.126}
$$

Note that (4.126) is also an LCT. We propose now to evaluate (4.126) through a direct application of the tableau of Fig. 4.1. Some elaboration is in order here. The tableaus in this paper have been derived with the implied assumption that the variables appearing in them are all scalars. This is not really necessary. Reviewing the derivations, one concludes that the tableaus are also valid under the following generalized interpretation:

1) The row (column) elements $(f_i, c_i, \beta_i, \delta_i, e_i, F_i, \ldots, \text{etc.})$ are column submatrices.
2) a_i, α_i, Ω are square matrices.
3) ω_i is still a scalar constant.
4) The tableau entries.

$$
\beta_i * \alpha_i = \delta_i \;; \qquad \beta_i * \omega_i \Omega = \delta_i
$$

should be interpreted as the following matrix products[10]

$$
\alpha_i \beta_i = \delta_i \;; \qquad \omega_i \Omega \beta_i = \delta_i.
$$

We introduce this generalization here with the immediate goal of evaluating (4.126). However, its importance transcends this immediate application: the generalization means that our basic DFT equation, namely (4.9), can now be interpreted as involving a block DFT matrix, that is, a block matrix whose (u, v) block is an arbitrary constant square matrix Ω multiplied by the scalar W^{uv}. All the DFT tableaus developed so far as well as the two final tableaus to be developed in this section are applicable to such a block DFT matrix. It is this generalization which elevates the tableaus from the rather theoretical realm of

---

10 This "modified interpretation" can be avoided by using row matrices instead of column matrices. Note in this context that the Ω matrix we shall be concerned with is symmetric.

fast algorithms for very low-order DFT's into the very practical realm of high-order, high-speed DFT algorithms. The specific way in which this is done is presented in detail in the next section.

We return now to the evaluation of (4.126). Applying Fig. 4.1, we get

$\hat{b}_0$	$\hat{b}_1$				$\hat{t}_1$	$\hat{t}_0$
1	1	$\hat{\beta}_1$	$*\hat{\alpha}_1=$	$\hat{\delta}_1$	1	1
1	-1	$\hat{\beta}_2$	$*\hat{\alpha}_2=$	$\hat{\delta}_2$	-1	1

$$(4.127)$$

$\frac{1}{2}\hat{a}_0$	$\frac{1}{2}\hat{a}_1$	
1	1	$\hat{\alpha}_1$
1	-1	$\hat{\alpha}_2$

$$(4.128)$$

Hence, (denoting $\gamma = 1/\sqrt{2}$),

$$\hat{\alpha}_1 = \frac{\Omega}{2}\begin{bmatrix} W^7+W^1 & W^5+W^3 \\ W^5+W^3 & W^7+W^1 \end{bmatrix} = \gamma\Omega\begin{bmatrix} 1 & -1 \\ -1 & 1 \end{bmatrix}; \quad \hat{\beta}_1 = \begin{bmatrix} f_7+f_1 \\ f_5+f_3 \end{bmatrix} = \begin{bmatrix} c_1 \\ c_3 \end{bmatrix}$$

$$\hat{\alpha}_2 = \frac{\Omega}{2}\begin{bmatrix} W^7-W^1 & W^5-W^3 \\ W^5-W^3 & W^7-W^1 \end{bmatrix} = i\gamma\Omega\begin{bmatrix} 1 & 1 \\ 1 & 1 \end{bmatrix}; \quad \hat{\beta}_2 = \begin{bmatrix} f_7-f_1 \\ f_5-f_3 \end{bmatrix} = \begin{bmatrix} c_7 \\ c_5 \end{bmatrix}$$

$$\hat{\delta}_1 = \hat{\alpha}_1\hat{\beta}_1 = \gamma\Omega\begin{bmatrix} 1 & -1 \\ -1 & 1 \end{bmatrix}\begin{bmatrix} c_1 \\ c_3 \end{bmatrix} = \gamma\Omega\begin{bmatrix} c_1-c_3 \\ c_3-c_1 \end{bmatrix} = \gamma\Omega\begin{bmatrix} \beta_1 \\ -\beta_1 \end{bmatrix} = \begin{bmatrix} \delta_1 \\ -\delta_1 \end{bmatrix}$$

$$\hat{\delta}_2 = \hat{\alpha}_2\hat{\beta}_2 = i\gamma\Omega\begin{bmatrix} 1 & 1 \\ 1 & 1 \end{bmatrix}\begin{bmatrix} c_7 \\ c_5 \end{bmatrix} = i\gamma\Omega\begin{bmatrix} c_5+c_7 \\ c_5+c_7 \end{bmatrix} = i\gamma\Omega\begin{bmatrix} \beta_7 \\ \beta_7 \end{bmatrix} = \begin{bmatrix} \delta_7 \\ \delta_7 \end{bmatrix}$$

$$\begin{bmatrix} F_7' \\ F_5' \end{bmatrix} = \hat{t}_1 = \hat{\delta}_1 - \hat{\delta}_2 = \begin{bmatrix} \delta_1-\delta_7 \\ -(\delta_1+\delta_7) \end{bmatrix} = \begin{bmatrix} g_7 \\ -g_1 \end{bmatrix}; \quad \begin{bmatrix} F_1' \\ F_3' \end{bmatrix} = \hat{t}_0 = \hat{\delta}_1 + \hat{\delta}_2 = \begin{bmatrix} \delta_1+\delta_7 \\ -(\delta_1-\delta_7) \end{bmatrix}$$

$$= \begin{bmatrix} g_1 \\ -g_7 \end{bmatrix}.$$

We turn now to the realization of the remaining parts of (4.123)

$$r=1;5$$

$$F_r = F_r' + \Omega[\underbrace{(f_0-f_4)}_{c_4}+i\underbrace{(f_6-f_2)}_{c_6}] = F_r' + \Omega(\underbrace{c_4+ic_6}_{\beta_4}) = F_r' + \underbrace{\Omega\beta_4}_{g_4} = F_r' + g_4$$

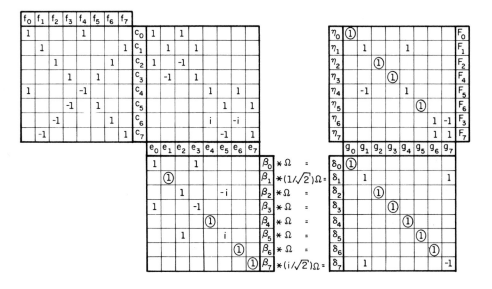

Fig. 4.14. Algorithm for DFT of order 8   N = 8   (8 M; 26 A)

$r = 3; 7$

$$F_r = F'_r + \Omega(\underbrace{c_4 - ic_6}_{\beta_6}) = F'_r + g_6$$

$$F_6 = \Omega[\underbrace{(f_0 + f_4)}_{c_0} - \underbrace{(f_2 + f_6)}_{c_2} - i\underbrace{(f_7 - f_1)}_{c_7} + i\underbrace{(f_5 - f_3)}_{c_5}] = \Omega[\underbrace{(c_0 - c_2)}_{e_2} + i\underbrace{(c_5 - c_7)}_{e_5}]$$

$$F_6 = \Omega(\underbrace{e_2 + ie_5}_{\beta_5}) = \Omega\beta_5 = g_5$$

$$F_2 = \Omega(\underbrace{e_2 - ie_5}_{\beta_2}) = \Omega\beta_2 = g_2$$

$$F_4 = \Omega[\underbrace{(f_0 + f_4)}_{c_0} + \underbrace{(f_6 + f_2)}_{c_2} - \underbrace{(f_7 + f_1)}_{c_1} - \underbrace{(f_5 + f_3)}_{c_3}] = \Omega[\underbrace{(c_0 + c_2)}_{e_0} - \underbrace{(c_1 + c_3)}_{e_3}]$$

$$= \Omega(\underbrace{e_0 - e_3}_{\beta_3}) = \Omega\beta_3 = g_3$$

$$F_0 = \Omega(\underbrace{e_0 + e_3}_{\beta_0}) = \Omega\beta_0 = g_0 .$$

This concludes the derivation.

4.6.2 DFT of Order 16 (Fig. 4.17)

The permutation is controlled by the following table:

ϱ	$\bar{0}$	$\bar{1}$	$\bar{2}$	$\bar{3}$	0	1	2	3
$r = 3^{\varrho} \bmod 16$	–	–	–	–	1	3	9	11
$r = -3^{\varrho} \bmod 16$	15	13	7	5	–	–	–	–

$$(4.129)$$

Applying this, we get the following E matrix:

$$
\begin{array}{c}
s \to \\
\sigma \to
\end{array}
\begin{array}{cccc|cccc|cccc|cccc}
0 & 4 & 8 & 12 & 2 & 6 & 10 & 14 & 15 & 13 & 7 & 5 & 1 & 3 & 9 & 11 \\
(0) & (4) & (8) & (12) & (2) & (6) & (10) & (14) & \bar{0} & \bar{1} & \bar{2} & \bar{3} & 0 & 1 & 2 & 3
\end{array}
$$

$$
E=
\left[
\begin{array}{cccc|cccc|cccc|cccc}
0 & 0 & 0 & 0 & 0 & 0 & 0 & 0 & 0 & 0 & 0 & 0 & 0 & 0 & 0 & 0 \\
0 & 0 & 0 & 0 & 8 & 8 & 8 & 8 & 12 & 4 & 12 & 4 & 4 & 12 & 4 & 12 \\
0 & 0 & 0 & 0 & 0 & 0 & 0 & 0 & 8 & 8 & 8 & 8 & 8 & 8 & 8 & 8 \\
0 & 0 & 0 & 0 & 8 & 8 & 8 & 8 & 4 & 12 & 4 & 12 & 12 & 4 & 12 & 4 \\
\hline
0 & 8 & 0 & 8 & 4 & 12 & 4 & 12 & 14 & 10 & 14 & 10 & 2 & 6 & 2 & 6 \\
0 & 8 & 0 & 8 & 12 & 4 & 12 & 4 & 10 & 14 & 10 & 14 & 6 & 2 & 6 & 2 \\
0 & 8 & 0 & 8 & 4 & 12 & 4 & 12 & 6 & 2 & 6 & 2 & 10 & 14 & 10 & 14 \\
0 & 8 & 0 & 8 & 12 & 4 & 12 & 4 & 2 & 6 & 2 & 6 & 14 & 10 & 14 & 10 \\
\hline
0 & 12 & 8 & 4 & 14 & 10 & 6 & 2 & 1 & 3 & 9 & 11 & 15 & 13 & 7 & 5 \\
0 & 4 & 8 & 12 & 10 & 14 & 2 & 6 & 3 & 9 & 11 & 1 & 13 & 7 & 5 & 15 \\
0 & 12 & 8 & 4 & 14 & 10 & 6 & 2 & 9 & 11 & 1 & 3 & 7 & 5 & 15 & 13 \\
0 & 4 & 8 & 12 & 10 & 14 & 2 & 6 & 11 & 1 & 3 & 9 & 5 & 15 & 13 & 7 \\
\hline
0 & 4 & 8 & 12 & 2 & 6 & 10 & 14 & 15 & 13 & 7 & 5 & 1 & 3 & 9 & 11 \\
0 & 12 & 8 & 4 & 6 & 2 & 14 & 10 & 13 & 7 & 5 & 15 & 3 & 9 & 11 & 1 \\
0 & 4 & 8 & 12 & 2 & 6 & 10 & 14 & 7 & 5 & 15 & 13 & 9 & 11 & 1 & 3 \\
0 & 12 & 8 & 4 & 6 & 2 & 14 & 10 & 5 & 15 & 13 & 7 & 11 & 1 & 3 & 9
\end{array}
\right]
\begin{array}{cc}
(0) & 0 \\
(4) & 4 \\
(8) & 8 \\
(12) & 12 \\
(2) & 2 \\
(6) & 6 \\
(10) & 10 \\
(14) & 14 \\
\bar{0} & 15 \\
\bar{1} & 13 \\
\bar{2} & 7 \\
\bar{3} & 5 \\
0 & 1 \\
1 & 3 \\
2 & 9 \\
3 & 11 \\
\varrho^{\uparrow} & r^{\uparrow}
\end{array}
\qquad (4.130)
$$

In view of the complexity of the present case, we derive the tableau in two stages. F'_i, the contribution of the LC submatrices, is considered separately in the intermediate tableau of Fig. 4.15 which is then copied into the final tableau of Fig. 4.17.

Following the previous case, we write down explicitly the LC part

$$
\left.\begin{array}{c}
\hat{t}_1 \left\{ \begin{bmatrix} F'_{15} \\ F'_{13} \\ F'_7 \\ F'_5 \end{bmatrix} \right. \\
\hat{t}_0 \left\{ \begin{bmatrix} F'_1 \\ F'_3 \\ F'_9 \\ F'_{11} \end{bmatrix} \right.
\end{array}\right]
=
\left[
\begin{array}{cccc|cccc}
\multicolumn{4}{c}{\hat{a}_1} & \multicolumn{4}{c}{\hat{a}_0} \\
w_1 & w_3 & w_9 & w_{11} & w_{15} & w_{13} & w_7 & w_5 \\
w_3 & w_9 & w_{11} & w_1 & w_{13} & w_7 & w_5 & w_{15} \\
w_9 & w_{11} & w_1 & w_3 & w_7 & w_5 & w_{15} & w_{13} \\
w_{11} & w_1 & w_3 & w_9 & w_5 & w_{15} & w_{13} & w_7 \\
\hline
w_{15} & w_{13} & w_7 & w_5 & w_1 & w_3 & w_9 & w_{11} \\
w_{13} & w_7 & w_5 & w_{15} & w_3 & w_9 & w_{11} & w_1 \\
w_7 & w_5 & w_{15} & w_{13} & w_9 & w_{11} & w_1 & w_3 \\
w_5 & w_{15} & w_{13} & w_7 & w_{11} & w_1 & w_3 & w_9
\end{array}
\right]
\left.\begin{array}{c}
\begin{bmatrix} f_{15} \\ f_{13} \\ f_7 \\ f_5 \end{bmatrix} \right\} \hat{b}_0 \\
\begin{bmatrix} f_1 \\ f_3 \\ f_9 \\ f_{11} \end{bmatrix} \right\} \hat{b}_1
\end{array}
,
\qquad (4.131)
$$

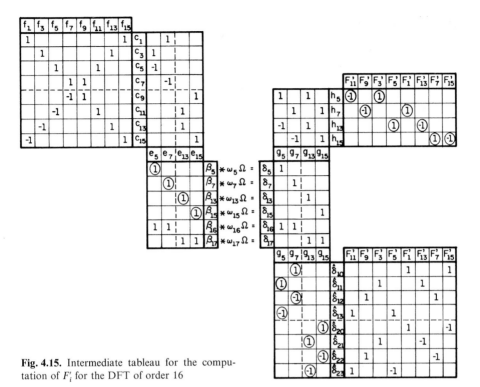

Fig. 4.15. Intermediate tableau for the computation of F_i' for the DFT of order 16

that is,

$$\begin{bmatrix} \hat{t}_1 \\ \hat{t}_0 \end{bmatrix} = \begin{bmatrix} \hat{a}_1 & \hat{a}_0 \\ \hat{a}_0 & \hat{a}_1 \end{bmatrix} \begin{bmatrix} \hat{b}_0 \\ \hat{b}_1 \end{bmatrix}. \tag{4.132}$$

We handle (4.132) via the tableaus (4.127, 128). It is obvious from (4.128) that $\hat{\alpha}_1, \hat{\alpha}_2$ will be LC matrices. Hence, it is sufficient to compute their first rows only. We express these in terms of the basic angle

$$\theta = \frac{360°}{16} = 22.5° \tag{4.133}$$

$$[\text{First row of } \hat{\alpha}_1] = \Omega \begin{bmatrix} \cos\theta & \sin\theta & -\cos\theta & -\sin\theta \end{bmatrix} \tag{4.134}$$

$$[\text{First row of } \hat{\alpha}_2] = i\Omega \begin{bmatrix} \sin\theta & \cos\theta & -\sin\theta & -\cos\theta \end{bmatrix}. \tag{4.135}$$

Turning to (4.127), we get

$$\hat{\beta}_1 = \begin{bmatrix} f_{15}+f_1 \\ f_{13}+f_3 \\ f_7 +f_9 \\ f_5 +f_{11} \end{bmatrix} = \begin{bmatrix} c_1 \\ c_3 \\ c_7 \\ c_5 \end{bmatrix} \quad ; \quad \hat{\beta}_2 = \begin{bmatrix} f_{15}-f_1 \\ f_{13}-f_3 \\ f_7 -f_9 \\ f_5 -f_{11} \end{bmatrix} = \begin{bmatrix} c_{15} \\ c_{13} \\ -c_9 \\ -c_{11} \end{bmatrix}. \tag{4.136}$$

The next step in (4.127) calls for the evaluation of two LCT's of order 4, namely,

$$\hat{\delta}_i = \hat{\alpha}_i \hat{\beta}_i \quad (i = 1, 2). \tag{4.137}$$

We adopt now the following notation for the components of $\hat{\delta}_i$:

$$\hat{\delta}_i = \begin{bmatrix} \hat{\delta}_{i0} \\ \hat{\delta}_{i1} \\ \hat{\delta}_{i2} \\ \hat{\delta}_{i3} \end{bmatrix}. \tag{4.138}$$

These will be determined now by applying the tableau of Fig. 4.2.

Implementation of $\hat{\delta}_1 = \hat{\alpha}_1 \hat{\beta}_1$

$-\dfrac{\Omega}{4}\sin\theta$	1			1
$-\dfrac{\Omega}{4}\cos\theta$		1	1	
$\dfrac{\Omega}{4}\sin\theta$	1			-1
$\dfrac{\Omega}{4}\cos\theta$		1	-1	

$$0 \quad 0 \quad -\frac{\Omega}{2}\cos\theta \quad -\frac{\Omega}{2}\sin\theta$$

1	1		0
1	-1		0
	①		$-\dfrac{\Omega}{2}\cos\theta$
		①	$-\dfrac{\Omega}{2}\sin\theta$
	1	1	$-\dfrac{\Omega}{2}(\cos\theta+\sin\theta)$

$$\left. \begin{array}{l} = \varepsilon_5 \Omega; \, \omega_5 = 2\varepsilon_{16} \\ = -(\cos\theta + \sin\theta) \\ = \varepsilon_7 \Omega; \, \omega_7 = 2(\varepsilon_7 - \varepsilon_5) \\ = (\cos\theta - \sin\theta) \\ = \varepsilon_{16}\Omega; \, \omega_{16} = -2\varepsilon_7 \\ = \sin\theta \end{array} \right\} \tag{4.139}$$

Note that the vanishing of the first two multipliers means that only a portion of Fig. 4.2 is required, namely, the portion involving the rows of $\beta_3, \beta_4, \beta_5$. This is copied into rows of $\beta_5, \beta_7, \beta_{16}$, respectively, in Fig. 4.15 (hence the adopted ω

indices). Note further that Fig. 4.15 shows two alternative paths leading from g_i to F'_r. The upper path should be ignored for now. The preceding discussion has derived that portion of the tableau leading to $\hat{\delta}_{1j}$. We turn now to $\hat{\delta}_{2j}$.

Implementation of $\hat{\delta}_2 = \hat{\alpha}_2\hat{\beta}_2$

$-\dfrac{i\Omega}{4}\cos\theta$	1		1		
$-\dfrac{i\Omega}{4}\sin\theta$		1	1		
$\dfrac{i\Omega}{4}\cos\theta$	1		-1		
$\dfrac{i\Omega}{4}\sin\theta$		1	-1		

0	0	$-\dfrac{i\Omega}{2}\sin\theta$	$-\dfrac{i\Omega}{2}\cos\theta$

1	1		0
1	-1		0
		①	$-\dfrac{i\Omega}{2}\sin\theta$
			①
	1	1	

$$-\dfrac{i\Omega}{2}\sin\theta \;=\varepsilon_{13}\Omega;\;\omega_{13}=2\varepsilon_{17}$$
$$=-i(\sin\theta+\cos\theta)$$
$$-\dfrac{i\Omega}{2}\cos\theta \;=\varepsilon_{15}\Omega;\;\omega_{15}=2(\varepsilon_{15}-\varepsilon_{13})$$
$$=i(\sin\theta-\cos\theta)$$
$$-\dfrac{i\Omega}{2}(\sin\theta+\cos\theta) \;=\varepsilon_{17}\Omega;\;\omega_{17}=-2\varepsilon_{15}$$
$$=i\cos\theta. \tag{4.140}$$

The situation here is very similar to the $\hat{\delta}_1$ case. The rows of $\beta_3, \beta_4, \beta_5$ of Fig. 4.2 are now copied into the rows of $\beta_{13}, \beta_{15}, \beta_{17}$, respectively, of Fig. 4.15.

With $\hat{\delta}_1, \hat{\delta}_2$ now available, we apply (4.127) to obtain $\hat{t}_1, \hat{t}_0$ as shown in the lower part of Fig. 4.15. The transition from g_i to F'_r is shown there requiring 8 additions. The upper part realizes the same transformation with only 4 additions and is the version copied into Fig. 4.17. The identity of the two paths can be easily verified by inspection. For example, the upper part prescribes $F'_1 = g_7 + g_{15}$ but so does the lower part. Verifying such agreements for all 8 outputs establishes the identity.

We implement now the remaining parts of (4.130). To bring out clearly the various symmetries we are exploiting here, we show an explicit form of the remaining part of the permuted DFT matrix in Fig. 4.16. It is shown here for

Y - Y' =

s →	0	4	8	12	2	6	10	14	15	13	7	5	1	3	9	11
σ →	(0)	(4)	(8)	(12)	(2)	(6)	(10)	(14)	$\bar{0}$	$\bar{1}$	$\bar{2}$	$\bar{3}$	0	1	2	3
	1	1	1	1	1	1	1	1	1	1	1	1	1	1	1	1
	1	1	1	1	-1	-1	-1	-1	i	-i	i	-i	-i	i	-i	i
	1	1	1	1	1	1	1	1	-1	-1	-1	-1	-1	-1	-1	-1
	1	1	1	1	-1	-1	-1	-1	-i	i	-i	i	i	-i	i	-i
	1	-1	1	-1	-i	i	-i	i	γ	-γ	γ	-γ	γ	-γ	γ	-γ
	1	-1	1	-1	i	-i	i	-i	-γ	γ	-γ	γ	-γ	γ	-γ	γ
	1	-1	1	-1	-i	i	-i	i	-γ	γ	-γ	γ	-γ	γ	-γ	γ
	1	-1	1	-1	i	-i	i	-i	γ	-γ	γ	-γ	γ	-γ	γ	-γ
	1	i	-1	-i	γ	-γ	-γ	γ								
	1	-i	-1	i	-γ	γ	γ	-γ								
	1	i	-1	-i	γ	-γ	-γ	γ								
	1	-i	-1	i	-γ	γ	γ	-γ								
	1	-i	-1	i	γ	-γ	-γ	γ								
	1	i	-1	-i	-γ	γ	γ	-γ								
	1	-i	-1	i	γ	-γ	-γ	γ								
	1	i	-1	-i	-γ	γ	γ	-γ								

+

0	4	8	12	2	6	10	14	15	13	7	5	1	3	9	11		
(0)	(4)	(8)	(12)	(2)	(6)	(10)	(14)	$\bar{0}$	$\bar{1}$	$\bar{2}$	$\bar{3}$	0	1	2	3	ρ	r
																(0)	0
																(4)	4
																(8)	8
																(12)	12
								iγ	iγ	iγ	iγ	-iγ	-iγ	-iγ	-iγ	(2)	2
								iγ	iγ	iγ	iγ	-iγ	-iγ	-iγ	-iγ	(6)	6
								-iγ	-iγ	-iγ	-iγ	iγ	iγ	iγ	iγ	(10)	10
								-iγ	-iγ	-iγ	-iγ	iγ	iγ	iγ	iγ	(14)	14
				iγ	iγ	-iγ	-iγ									$\bar{0}$	15
				iγ	iγ	-iγ	-iγ									$\bar{1}$	13
				iγ	iγ	-iγ	-iγ									$\bar{2}$	7
				iγ	iγ	-iγ	-iγ									$\bar{3}$	5
				-iγ	-iγ	iγ	iγ									0	1
				-iγ	-iγ	iγ	iγ									1	3
				-iγ	-iγ	iγ	iγ									2	9
				-iγ	-iγ	iγ	iγ									3	11

Fig. 4.16. *E* matrix for the computation of $F_i - F_i'$ for the DFT of order 16

convenience as the sum of two matrices and expressed in terms of the constant

$$\gamma = \frac{1}{\sqrt{2}}. \tag{4.141}$$

We define now

$$c_{14} = f_6 - f_{14}; \quad c_{12} = f_4 - f_{12}; \quad c_{10} = f_2 - f_{10}; \quad c_8 = f_0 - f_8$$

and proceed with the evaluation as follows:

r=1; 9

$$F_r - F'_r = \Omega\{\underbrace{(c_8 - ic_{12})}_{\beta_{12}} - \gamma[\underbrace{(c_{14} - c_{10})}_{e_{10}} + i\underbrace{(c_{14} + c_{10})}_{e_{14}}]\} = \underbrace{\Omega\beta_{12}}_{\delta_{12}} - i\gamma\Omega\underbrace{(e_{14} - ie_{10})}_{\beta_{10}}$$

$$= \underbrace{\delta_{12}}_{g_{12}} - i\gamma\Omega\underbrace{\beta_{10}}_{\delta_{10}} = g_{12} - \underbrace{\delta_{10}}_{g_{10}} = g_{12} - g_{10} = h_{12}$$

$$\therefore F_r = F'_r + h_{12}$$

r=5; 13

$$F_r = F'_r + \underbrace{(g_{12} + g_{10})}_{h_{10}} = F'_r + h_{10}$$

r=3; 11

$$F_r - F'_r = \Omega\{\underbrace{(c_8 + ic_{12})}_{\beta_8} + \gamma[\underbrace{(c_{14} - c_{10})}_{e_{10}} - i\underbrace{(c_{14} + c_{10})}_{e_{14}}]\} = \underbrace{\Omega\beta_8}_{\delta_8} - i\gamma\Omega\underbrace{(e_{14} + ie_{10})}_{\beta_{14}}$$

$$= \underbrace{\delta_8}_{g_8} - i\gamma\Omega\underbrace{\beta_{14}}_{\delta_{14}} = g_8 - \underbrace{\delta_{14}}_{g_{14}} = g_8 - g_{14} = h_8$$

$$F_r = F'_r + h_8$$

r=7; 15

$$F_r = F'_r + \underbrace{(g_8 + g_{14})}_{h_{14}} = F'_r + h_{14} \quad \text{--- ---}$$

We define now

$$c_0 = f_0 + f_8; \quad c_2 = f_2 + f_{10}; \quad c_4 = f_4 + f_{12}; \quad c_6 = f_6 + f_{14}$$

$$g_4 = \delta_4 = \Omega\beta_4 = \Omega(c_0 - c_4)$$

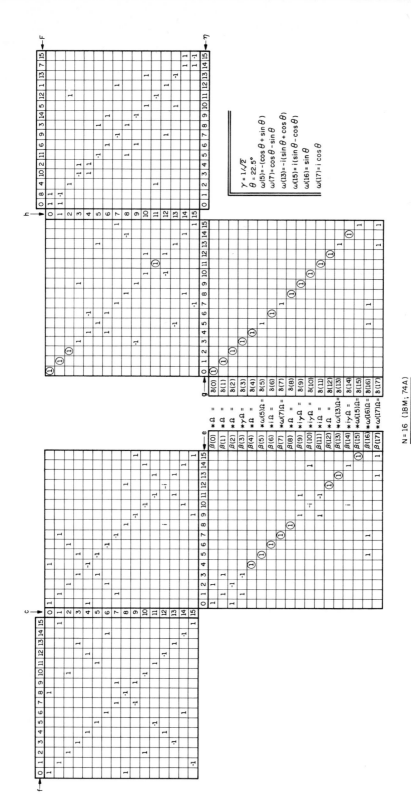

Fig. 4.17. Algorithm for DFT of order 16

$$N = 16 \quad (18M; 74A)$$

$\gamma = 1/\sqrt{2}$
$\theta = 22.5°$
$\omega(5) = -(\cos\theta + \sin\theta)$
$\omega(7) = \cos\theta - \sin\theta$
$\omega(13) = -i(\sin\theta + \cos\theta)$
$\omega(15) = i(\sin\theta - \cos\theta)$
$\omega(16) = \sin\theta$
$\omega(17) = i\cos\theta$

and turn to the next group of rows $(F_2, F_6, F_{10}, F_{14})$

$$F_2 = g_4 + \Omega\{-i\underbrace{(c_2 - c_6)}_{\beta_6} + \gamma[\underbrace{(c_7 + c_1)}_{e_1} - \underbrace{(c_5 + c_3)}_{e_3} + i\underbrace{(c_{15} - c_9)}_{e_9} + i\underbrace{(c_{13} - c_{11})}_{e_{11}}]\}$$

$$= g_4 - i\Omega\underbrace{\beta_6}_{\delta_6} + \gamma\Omega\underbrace{(e_1 - e_3)}_{\beta_3} + i\gamma\Omega\underbrace{(e_9 + e_{11})}_{\beta_9} = g_4 - \underbrace{\delta_6}_{g_6} + \gamma\Omega\underbrace{\beta_3}_{\delta_3} + i\gamma\Omega\underbrace{\beta_9}_{\delta_9}$$

$$= g_4 - g_6 + \underbrace{\delta_3}_{g_3} + \underbrace{\delta_9}_{g_9}$$

$$= \underbrace{(g_4 - g_6)}_{h_4} + \underbrace{(g_9 + g_3)}_{h_3} = h_4 + h_3 .$$

Noting which columns in Fig. 4.16 are associated with each of the four g_i's comprising F_2, we observe that F_6, F_{10}, F_{14} involve the same g_i's but with different sign combinations. Specifically,

$$F_{10} = (g_4 - g_6) - (g_9 + g_3) = h_4 - h_3$$

$$F_6 = \underbrace{(g_4 + g_6)}_{h_6} + \underbrace{(g_9 - g_3)}_{h_9} = h_6 + h_9$$

$$F_{14} = (g_4 + g_6) - (g_9 - g_3) = h_6 - h_9 .$$

Finally we consider the upper four rows

$$F_0 = \Omega[\underbrace{(c_0 + c_4)}_{e_0} + \underbrace{(c_2 + c_6)}_{e_2} + \underbrace{(c_7 + c_1)}_{e_1} + \underbrace{(c_5 + c_3)}_{e_3}] = \Omega[\underbrace{(e_0 + e_2)}_{\beta_0} + \underbrace{(e_1 + e_3)}_{\beta_1}]$$

$$= \underbrace{\Omega\beta_0}_{\delta_0} + \underbrace{\Omega\beta_1}_{\delta_1} = \underbrace{\delta_0}_{g_0} + \underbrace{\delta_1}_{g_1} = \underbrace{g_0}_{h_0} + \underbrace{g_1}_{h_1} = h_0 + h_1$$

$$\therefore F_8 = h_0 - h_1$$

$$F_4 = \Omega\{\underbrace{(e_0 - e_2)}_{\beta_2} + i[\underbrace{(c_{15} - c_9)}_{e_9} - \underbrace{(c_{13} - c_{11})}_{e_{11}}]\} = \underbrace{\Omega\beta_2}_{\delta_2} + i\Omega\underbrace{(e_9 - e_{11})}_{\beta_{11}}$$

$$= \underbrace{\delta_2}_{g_2} + i\Omega\underbrace{\beta_{11}}_{\delta_{11}}$$

$$= \underbrace{g_2}_{h_2} + \underbrace{\delta_{11}}_{g_{11}} = h_2 + \underbrace{g_{11}}_{h_{11}} = h_2 + h_{11}$$

$$\therefore F_{12} = h_2 - h_{11} .$$

This completes the derivation. Note that the size of this tableau imposes certain notational peculiarities. In particular, $\omega(i) = \omega_i$, $\beta(i) = \beta_i$, etc.

4.7 The Overall Algorithm[11]

At this point, we finally are in possession of DFT tableaus for all orders listed in (4.3). Our goal here is their integration into a DFT algorithm of order

$$N = \prod_{k=1}^{\kappa} N_k \quad (N_k\text{'s relatively prime}). \tag{4.142}$$

Our basic approach is based on the idea of a block DFT matrix introduced in the last section. Let us illustrate this approach with a simple example: suppose we want to implement the DFT of order $N = 120 = 8 \cdot 15$. Could we apply the tableau of order 8 to this task? We can certainly consider the original DFT matrix as a block matrix or order 8, with blocks of order 15. However, application of the 8th-order tableau is valid only if this block matrix displays the structure of a block DFT matrix of order 8. Simple examination reveals that this is definitely not the case. Still, this does not invalidate our approach. Conceivably, one could scramble the input and output vectors in such a way that the scrambled transformation matrix would in fact be an 8th-order block DFT matrix. If this were true, then we could indeed apply the 8th-order tableau (Fig. 4.14). In doing this, however, we would soon face another hurdle: the tableau calls for the evaluation of $\Omega\beta_i$ where Ω is now the 15th-order $(0, 0)$ block of the scrambled block DFT matrix. There are 8 such transformations in the tableau and to maintain the overall high computational efficiency, we must avoid direct matrix multiplication in their evaluation. An intriguing possibility now presents itself: is it possible to impose additional constraints on the initial scrambling so as to ensure that Ω turns out to be a block DFT matrix of (say) order 3? If this were the case, it would allow us to take advantage of the high efficiency of the DFT tableau of order 3 and thus we would end up implementing the DFT of order 120 by using the tableaus of orders 8, 3, 5.

We turn now to a more precise formulation of the above ideas, starting with the development of the necessary terminology. We regard the indices in (4.142) as reflecting the order of tableau realization. Thus, N_1 is the order of the first tableau realized, N_2 refers to the second tableau, etc. In our example

$$\kappa = 3; \quad N_1 = 8; \quad N_2 = 3; \quad N_3 = 5; \quad \therefore N = 120. \tag{4.143}$$

Let Ω_0 be the initial scrambled DFT matrix of order N. If we regard it as a block DFT matrix or order N_1, then each of its blocks is of order $v_1 = N/N_1$.

---

11 The basic idea underlying the developments in this section is commonly attributed to *Good* [4.10]. Here, we find it more convenient to avoid Good's explicit use of the Kronecker matrix product.

We denote the $(0,0)$ block by Ω_1. The next step in the realization is based on regarding Ω_1 as a block DFT matrix or order N_2. Its $(0,0)$ block which we denote Ω_2 is of order $v_2 = v_1/N_2$, etc. This leads us to the following definitions:

$$\left. \begin{aligned} v_i &= \prod_{k=i+1}^{\kappa} N_k \quad (0 \leq i < \kappa) \\ v_\kappa &= 1. \end{aligned} \right\} \tag{4.144}$$

Let $\Omega_0 (= Y)$ be the scrambled DFT matrix of order N. Then Ω_i is its upper-left submatrix of order v_i.

Note that (4.144) implies the existence of a sequence of these submatrices

$$\Omega_0, \Omega_1, \ldots, \Omega_\kappa.$$

The matrix order decreases to the right. Ω_0 on the left is of order N and is thus identical with the overall matrix. Ω_κ on the right is of order 1 and is thus the upper-left element of the overall matrix.

We can now formalize the set of sufficient scrambling constraints that would allow implementation of the above scheme:

1) Ω_0 should be a block DFT matrix of order N_1, that is, a block matrix whose (m, n) block is the matrix Ω_1 multiplied by W^{mn} where $W = \exp(-i2\pi/N_1)$ [see (4.9)]. This will allow the implementation of phase 1 of the N_1 tableau. Phase 2 calls for the determination of $(\omega_i \Omega_1)\beta_i$. Hence,

2) Ω_1 should be a block DFT matrix of order N_2. We apply phase 1 of the N_2 tableau and are led, in phase 2, to a transformation involving Ω_2. Hence,

3) Ω_2 should be a block DFT matrix or order N_3.

$\vdots$

κ) $\Omega_{\kappa-1}$ should be a DFT matrix of order κ.

All this can be summarized as follows: An acceptable relabeling scheme would be one which satisfies the following set of κ constraints:

$$\left. \begin{aligned} &\text{Submatrix } \Omega_{k-1} \text{ of } Y \text{ should be a block DFT matrix or order } N_k. \\ &\text{This should hold for all } 1 \leq k \leq \kappa. \end{aligned} \right\} \tag{4.145}$$

While an algorithm following the above outline would be quite convenient to implement, it is not at all clear that a scrambling scheme that would simultaneously satisfy all κ constraints of (4.145) does, in fact, exist.

We proceed now to develop a relabeling scheme which comes very close to (4.145), adding only a minor complication to the above algorithm outline. We start with the standard DFT matrix $[(4.9)$ with $\Omega=1]^{12}$.

$$\hat{F}_u = \sum_{v=0}^{N-1} W^{uv}\hat{f}_v \quad (u=0, 1, \ldots, N-1). \tag{4.146}$$

Let

$$Q_m = \hat{F}_u; \quad q_n = \hat{f}_v, \tag{4.147}$$

where

$$u = \lambda(m); \quad v = \lambda(n) \quad (m, n = 0, 1, \ldots, N-1) \tag{4.148}$$

and the function λ is yet to be specified. This transforms (4.146) into

$$Q_m = \sum_{n=0}^{N-1} W^{\lambda(m)\lambda(n)}q_n = \sum_{n=0}^{N-1} Y_{mn}q_n \quad (m=0, 1, \ldots, N-1). \tag{4.149}$$

The function λ will be defined in terms of a modular representation [4.7] of u, v. In other words, the relabeling depends on the entities

$$\left.\begin{array}{l} u_k = u \bmod N_k \\ v_k = v \bmod N_k \end{array}\right\} \quad (k=1, 2, \ldots, \kappa). \tag{4.150}$$

According to the Chinese Remainder Theorem [4.7], these remainders uniquely determine any $0 \le (u, v) < N$. Hence, we may adopt the following representation for u, v:

$$\left.\begin{array}{l} u = (u_1, u_2, \ldots, u_\kappa) \\ v = (v_1, v_2, \ldots, v_\kappa) \end{array}\right\}. \tag{4.151}$$

The function λ is now defined in terms of the combination of (4.148, 151)

$$v = (v_1, v_2, \ldots, v_\kappa) = \lambda(n) \tag{4.152}$$

as follows:

$\lambda(n)$ should be such that as n follows the sequence $0, 1, \ldots, N-1$; each v_k should follow a periodic repetition of the sequence $0, 1, \ldots, N_k-1$, starting with $v_k=0$. v_k should be stepped with every v_k increment of n [see (4.144)]. This means that v_1 varies very slowly, v_2 varies faster, and so on, up to v_κ which varies in step with n.

---

12 We use here $\hat{F}_u, \hat{f}_v$ to distinguish these entities from the tableau variables F_u, f_v.

To illustrate this scrambling, consider our example (4.143) for which part of the index sequence would look as follows:

n	v
$\vdots$	$\vdots$
3	$48 = (0, 0, 3)$
4	$24 = (0, 0, 4)$
5	$40 = (0, 1, 0)$
6	$16 = (0, 1, 1)$
$\vdots$	$\vdots$
62	$12 = (4, 0, 2)$
63	$108 = (4, 0, 3)$
64	$84 = (4, 0, 4)$
65	$100 = (4, 1, 0)$
66	$76 = (4, 1, 1)$
$\vdots$	$\vdots$

One could use modular arithmetic subroutines to determine this sequence on a computer. Alternatively, it could be determined in a scheme using neither computers nor computations. All that is called for is some tedious writing down of repetitive sequences. We illustrate this method for our example (4.143) in Tables 4.2, 3. First we write down the sequence $0, 1, ..., N-1$ as shown in Table 4.2 under the heading v. Then we write down next to this column, under the heading v_k, a periodic repetition of the sequence $0, 1, ..., N_k - 1$ and repeat this for all $1 \leq k \leq \kappa$. This yields representation (4.151) for all v's. The particular arrangement in Table 4.2 saves some writing and is convenient for the next step which is just a recordering of Table 4.2 in the desired sequence.

To simplify the process, we let Table 4.2 determine the order in which Table 4.3 is being filled in. For example, the first v values copied from Table 4.2 into Table 4.3 are $0, 40, 80, 8, 48, 88, 16, 56, ...,$ etc. Table 4.3 then prescribes the index sequence of the scrambled input vector for (4.143), namely,

$$\hat{f}_0, \hat{f}_{96}, \hat{f}_{72}, ..., \hat{f}_{104}, \hat{f}_{105}, ..., \hat{f}_{89}, \hat{f}_{90}, ..., \hat{f}_{119}.$$

To facilitate the analysis of the adopted scrambling, we introduce now E, the exponent matrix of Y. Recall that (4.149) implies

$$Y_{mn} = W^{\lambda(m)\lambda(n)}. \tag{4.153}$$

Table 4.2. Modular representation of v in example (4.143); $N_1=8$; $N_2=3$; $N_3=5$

v_1	v	v_2	v_3	v	v_2	v_3	v	v_2	v_3	v	v_2	v_3	v	v_2	v_3
0	0	0	0	8	2	3	16	1	1	24	0	4	32	2	2
1	1	1	1	9	0	4	17	2	2	25	1	0	33	0	3
2	2	2	2	10	1	0	18	0	3	26	2	1	34	1	4
3	3	0	3	11	2	1	19	1	4	27	0	2	35	2	0
4	4	1	4	12	0	2	20	2	0	28	1	3	36	0	1
5	5	2	0	13	1	3	21	0	1	29	2	4	37	1	2
6	6	0	1	14	2	4	22	1	2	30	0	0	38	2	3
7	7	1	2	15	0	0	23	2	3	31	1	1	39	0	4
0	40	1	0	48	0	3	56	2	1	64	1	4	72	0	2
1	41	2	1	49	1	4	57	0	2	65	2	0	73	1	3
2	42	0	2	50	2	0	58	1	3	66	0	1	74	2	4
3	43	1	3	51	0	1	59	2	4	67	1	2	75	0	0
4	44	2	4	52	1	2	60	0	0	68	2	3	76	1	1
5	45	0	0	53	2	3	61	1	1	69	0	4	77	2	2
6	46	1	1	54	0	4	62	2	2	70	1	0	78	0	3
7	47	2	2	55	1	0	63	0	3	71	2	1	79	1	4
0	80	2	0	88	1	3	96	0	1	104	2	4	112	1	2
1	81	0	1	89	2	4	97	1	2	105	0	0	113	2	3
2	82	1	2	90	0	0	98	2	3	106	1	1	114	0	4
3	83	2	3	91	1	1	99	0	4	107	2	2	115	1	0
4	84	0	4	92	2	2	100	1	0	108	0	3	116	2	1
5	85	1	0	93	0	3	101	2	1	109	1	4	117	0	2
6	86	2	1	94	1	4	102	0	2	110	2	0	118	1	3
7	87	0	2	95	2	0	103	1	3	111	0	1	119	2	4

Table 4.3. Scrambling for example (4.143); n vs $v=(v_1,v_2,v_3)$

v_1								v_2	v_3
0	1	2	3	4	5	6	7		
0	105	90	75	60	45	30	15	0	0
96	81	66	51	36	21	6	111	0	1
72	57	42	27	12	117	102	87	0	2
48	33	18	3	108	93	78	63	0	3
24	9	114	99	84	69	54	39	0	4
40	25	10	115	100	85	70	55	1	0
16	1	106	91	76	61	46	31	1	1
112	97	82	67	52	37	22	7	1	2
88	73	58	43	28	13	118	103	1	3
64	49	34	19	4	109	94	79	1	4
80	65	50	35	20	5	110	95	2	0
56	41	26	11	116	101	86	71	2	1
32	17	2	107	92	77	62	47	2	2
8	113	98	83	68	53	38	23	2	3
104	89	74	59	44	29	14	119	2	4

Hence we define the exponent matrix E as follows:

$$E_{mn} = \lambda(m)\lambda(n) = uv . \tag{4.154}$$

Similarly, paralleling the Ω_k submatrices of Y we define $\mathscr{E}_k$ as the upper-left v_k-order submatrix of E.

Consider now the distribution of the arguments u_k, v_k in $\mathscr{E}_{k-1}$. The scrambling prescribes that, starting with the value zero at the upper-left corner, u_k should be increased by one every v_k rows. Similarly, v_k should increase by one every v_k columns. This, then defines a subdivision of $\mathscr{E}_{k-1}$ into submatrices of order v_k. $\mathscr{E}_{k-1}$ can now be regarded as a block matrix of order $N_k (= v_{k-1}/v_k)$, whose (r, s) block is characterized by

$$u_k = r ; \quad v_k = s . \tag{4.155}$$

The $(0,0)$ block of this block matrix is obviously identical with $\mathscr{E}_k$. Let us pick now an element in an arbitrary position in $\mathscr{E}_k$. Its u, v will have the form

$$\left.\begin{array}{l} u = (0, \ldots, 0, u_{k+1}, u_{k+2}, \ldots, u_\kappa) \\ v = (0, \ldots, 0, v_{k+1}, v_{k+2}, \ldots, v_\kappa) \end{array}\right\} . \tag{4.156}$$

Now, the adopted scrambling scheme guarantees that all scalar elements of $\mathscr{E}_{k-1}$ occupying the same identical position in the other blocks of $\mathscr{E}_{k-1}$ have u_i, v_i which differ from (4.156) only in u_k, v_k and these satisfy (4.155). This means that the difference between an element in block (r, s) of $\mathscr{E}_{k-1}$ and the corresponding element in $\mathscr{E}_k$ [block $(0,0)$] is just ([4.7], (4.154))

$$\Delta E(r, s) = (0, \ldots, 0, rs \bmod N_k, 0, \ldots, 0) . \tag{4.157}$$

$$\uparrow$$
$$k\text{th position}$$

We notice the important fact that the difference is independent of the specific common location of the two paired elements in their respective blocks. Hence, the difference is constant throughout the (r, s) block. In other words, the (r, s) block of $\mathscr{E}_{k-1}$ could be generated from $\mathscr{E}_k$ simply by adding the constant $\Delta E(r, s)$ to all its elements.

We are interested in an explicit expression for this important constant. Considering its implicit formulation (4.157), we conclude that it must be some integral multiple of n_k where

$$n_i = \prod_{\substack{k=1 \\ k \neq i}}^\kappa N_k = \frac{N}{N_i} . \tag{4.158}$$

Specifically, we must have

$$\Delta E(r, s) = \eta(r, s) n_k, \tag{4.159}$$

where the integer η satisfies

$$\eta < N_k \tag{4.160}$$

$$n_k \eta - rs \equiv 0 \pmod{N_k}. \tag{4.161}$$

Equation (4.161) is a linear congruence for the unknown η for which there is an explicit solution [4.8], namely,

$$\eta = (rs') \bmod N_k, \tag{4.162}$$

where[13]

$$s' = (s\zeta_k) \bmod N_k \tag{4.163}$$

$$\zeta_k = n_k^{\phi(N_k)-1} \bmod N_k. \tag{4.164}$$

The result we have established for the submatrices of $\mathscr{E}_{k-1}$ translates as follows for the submatrices of Ω_{k-1}: The (r, s) block of Ω_{k-1} is just $W^{\Delta E(r,s)} \Omega_k$. Applying (4.159), we find that

$$W^{\Delta E(r,s)} = \exp{-i \frac{2\pi}{N_k} \eta(r, s)}. \tag{4.165}$$

Denoting now

$$W_k = \exp{-i \frac{2\pi}{N_k}}, \tag{4.166}$$

we get

$$W^{\Delta E(r,s)} = W_k^{\eta(r,s)} \tag{4.167}$$

so that the (r, s) block of the N_kth order block matrix Ω_{k-1} is

$$W_k^{\eta(r,s)} \Omega_k = W_k^{(rs') \bmod N_k} \Omega_k = W_k^{rs'} \Omega_k.$$

This is very similar to the (r, s) block of the block DFT matrix (4.9) of order N_k with $\Omega = \Omega_k$. The only difference is that s' has now replaced s. This, however, is a

---

13 $\phi(n)$ is the Euler Totient function defined as the number of integers not exceeding, and relatively prime to, n.

trivial difference involving only column permutations. To prove this, it is sufficient to show that there is a one-to-one correspondence between s and s' so that when s goes through the values $0, 1, ..., N_k - 1$, s' goes through a permutation of them. This will, indeed, be the case if ζ_k [in (4.163)] is relatively prime to N_k. But this is guaranteed by (4.164) since n_k is relatively prime to N_k [see (4.158)].

We conclude that a trivial modification of the DFT tableau of order N_k will evaluate the transformation effected by Ω_{k-1}. Specifically, we should modify the input square of the N_k tableau by permuting the f_i rows/columns so that the i sequence would be identical with the s' sequence [instead of the natural number sequence (s) of the unmodified tableau]. We refer to such a tableau as a "modified tableau" and use this term from now on, only with this restricted, precise, meaning.

We illustrate now the tableau modification with $k = 1$ in our example (4.143)

$$n_1 = N_2 N_3 = 15 ; \quad \phi(N_1) = \phi(8) = 4$$
$$\zeta_1 = 15^{4-1} \bmod 8 = (-1)^3 \bmod 8 = 7 .$$

Hence

s	0	1	2	3	4	5	6	7
$s' = (7s) \bmod 8$	0	7	6	5	4	3	2	1

(4.168)

The modified input square called for by (4.168) is shown in Fig. 4.18. Note that, in this case, the modification involves only sign changes.

The result we have established may be summarized as follows:

Submatrix Ω_{k-1} of Y is a column permutation of the block DFT matrix of order N_k (4.9) with $\Omega = \Omega_k$. This is valid for all $1 \leq k \leq \kappa$. $\left.\right\}$ (4.169)

f_0	f_1	f_2	f_3	f_4	f_5	f_6	f_7	
1				1				c_0
	1						1	c_1
		1				1		c_2
			1		1			c_3
1				-1				c_4
	1					-1		c_5
		1					-1	c_6
	1						-1	c_7

Fig. 4.18. Input square of the 8th-order modified tableau for example (4.143)

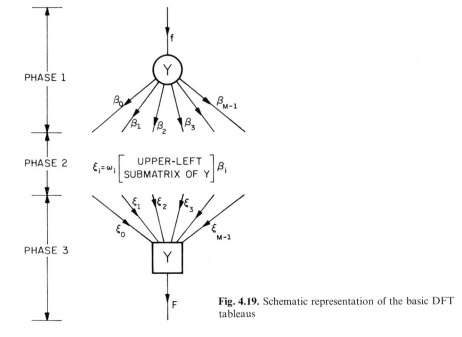

PHASE 1

PHASE 2

PHASE 3

F

Fig. 4.19. Schematic representation of the basic DFT tableaus

This differs from (4.145) in the column permutation. But, as we have just pointed out, this only means that in the algorithm for order N (4.142), the standard tableaus should be replaced by the modified tableaus. This is the minor complication referred to earlier.

We conclude this section with an important graphical representation of the overall algorithm. This is based on the symbolic representation of the tableaus shown in Fig. 4.19. We phrase this in terms of their generalized, block matrix, interpretation: let the tableau variables $f_i, \beta_i, \delta_i, F_i, \ldots$, etc., represent m-dimensional vectors. Then the tableau of order N is applicable to any matrix transformation of order mN in which the transforming matrix Y, when regarded as a block matrix of order N, is the Nth-order DFT matrix (4.9). In this case, the Ω parameter of the tableau is the mth order upper-left submatrix of Y.

Examination of the tableaus reveals that they all consist of three parts corresponding to three distinct phases of the algorithms they describe. In phase 1, the mN-dimensional input vector f is operated upon to yield the m-dimensional β_i vectors. In phase 2, a scalar multiple of the Ω submatrix transforms β_i into ξ_i according to (4.86) $[\xi_i = (\omega_i \Omega)\beta_i; i = 0, 1, \ldots, M-1]$[14]. M is the number of multiplications appearing in the tableau designation. Obviously, this is also the number of β_i vectors generated in phase 1. Finally, in phase 3, the m-dimensional ξ_i vectors are operated upon to yield the mN-dimensional

---

14 For most i values, $\xi_i = \delta_i$ of the tableaus but not for all of them. For example, the $N = 5$ tableau implies $\xi_0 = F_0$, $\delta_1 = \xi_1 + \xi_0$.

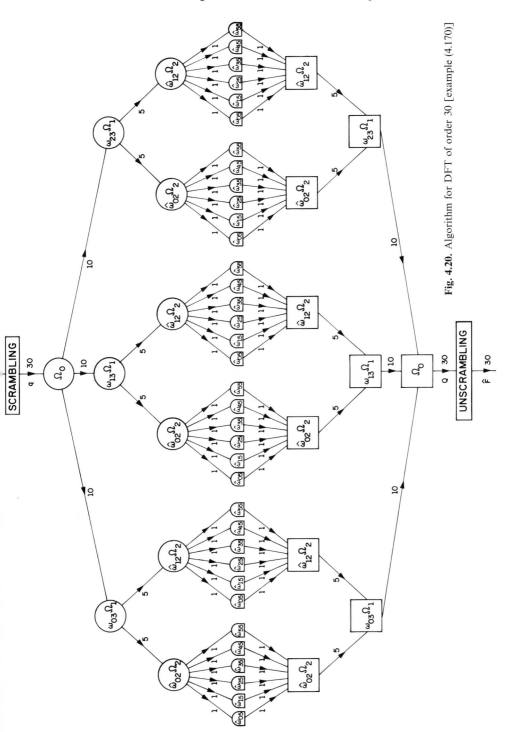

Fig. 4.20. Algorithm for DFT of order 30 [example (4.170)]

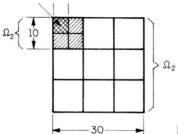

Fig. 4.21. Subdivision of the Y matrix for example (4.170)

output vector F. The three phases are represented schematically in Fig. 4.19. The conventions adopted here are as follows: All lines represent vectors. We may attach an integer to a line to indicate the dimensionality of the vector it represents (see Fig. 4.20). Phase 1 is represented by a circle, phase 3 by a square. In either case, the symbol inside designates the matrix transforming the "circle input" to the "square output". Note that the only arithmetic operations involved inside either the circle or the square are additions and subtractions.

The overall algorithm appears in Fig. 4.20. We show here a specific example, namely:

$$\kappa = 3; \quad N_1 = 3; \quad N_2 = 2; \quad N_3 = 5; \quad \therefore N = 30. \tag{4.170}$$

The modification for any other case is quite simple as will be indicated later on. The first step of the algorithm is the scrambling of the input vector $\hat{f}$ to yield the vector q. This and the concurrent scrambling of the output $\hat{F}$, transform the DFT matrix into the Y matrix of (4.153). Y is now subdivided as indicated schematically in Fig. 4.21 (the indicated "measurements" refer to the number of rows/columns). The next step is to regard $Y(=\Omega_0)$ as a block third-order matrix ($N_1 = 3$) and apply the third-order modified DFT tableau with the tableau's Ω identified with Ω_1 of Fig. 4.21. Phase 1 is shown at the top of Fig. 4.20 (circled Ω_0), phase 3 is at the bottom (Ω_0 in a square), and phase 2 is all the region in between. Phase 1 requires separation of the input vector into its three "components". This is implemented in the following straightforward manner:

$$f_0 = \begin{bmatrix} q_0 \\ \vdots \\ q_9 \end{bmatrix}; \quad f_1 = \begin{bmatrix} q_{10} \\ \vdots \\ q_{19} \end{bmatrix}; \quad f_2 = \begin{bmatrix} q_{20} \\ \vdots \\ q_{29} \end{bmatrix}. \tag{4.171}$$

The outputs of phase 1 are the three β_i vectors of dimensionality 10. Before considering phase 2, we have to introduce the generalized notation we use there for the ω_i's. ω_i for the DFT tableau of order N is referred to now as $\omega_{i,N}$. ($\hat{\omega}_{i,N}$, also appearing in Fig. 4.20, will be defined later). These constants are either explicitly listed in the tableaus or are inferred from the convention that every β_i

Table 4.4. The DFT tableau multipliers $(\omega_{k,N})$

k \\ N	2	3	4	5	7	8	9	16
0	1	1	1	1	1	1	1	1
1	1	− 1.5	1	− 1.25	− 1.1666667E − 0	7.0710678E − 1	0.5	1
2		i8.6602540E − 1	1	− 5.5901699E − 1	5.5854267E − 2	1	− 1.7364818E − 1	1
3			i	i1.5388418E − 0	7.3430220E − 1	1	0.5	7.0710678E − 1
4				− i3.6327126E − 1	− i8.7484229E − 1	1	9.3969262E − 1	1
5				− i5.8778525E − 1	− i5.3396936E − 1	1	− i3.4202014E − 1	− 1.3065630E − 0
6					− i4.4095855E − 1	1	− i8.6602540E − 1	i
7					7.9015647E − 1	i7.0710678E − 1	− i9.8480775E − 1	5.4119610E − 1
8					− i3.4087293E − 1		− i8.6602540E − 1	1
9							7.6604444E − 1	i7.0710678E − 1
10							− i6.4278761E − 1	i7.0710678E − 1
11								i
12								1
13								− i1.3065630E − 0
14								i7.0710678E − 1
15								− i5.4119610E − 1
16								3.8268343E − 1
17								i9.2387953E − 1

Note: Numbers shown in 3 digits or less are exact. All other numbers are in Fortran E-format ($3.4E − 1 = 3.4 \times 10^{-1}$).

is multiplied by $(\omega_i\Omega)$. For example, the DFT tableau of order 5 states explicitly $\omega_{1,5} = -5/4$. Implicitly, we infer $\omega_{0,5} = 1$. All the ω_{ij}'s are tabulated in Table 4.4 for convenience. The values there have been computed on a 10-digit calculator. For more precise values, one should refer to the exact expressions in the tableaus.

Returning now to Fig. 4.20, we find that phase 2 of the third-order tableau requires the evaluation of $\xi_i = (\omega_{i3}\Omega_1)\beta_i$. Here we apply again result (4.169) which implies (with $k=2$) that the transformation $(\omega_{i3}\Omega_1)\beta_i$ could be evaluated with the modified second-order DFT tableau [with the tableau's Ω identified with $(\omega_{i3}\Omega_2)$ (see Fig. 4.21)]. Each of the three 10-dimensional β_i's is therefore shown in Fig. 4.20 as the input to phase 1 of a second-order tableau. In each of these applications of the second-order tableau, phase 1 yields a pair of 5-dimensional β vectors. Consider now the specific 5-dimensional β vector on the extreme left of Fig. 4.20. It has been generated by phase 1 of the tableau implementing the transformation based on the matrix $(\omega_{03}\Omega_1)$. Therefore, phase 2 calls for its transformation by the matrix $\omega_{02}(\omega_{03}\Omega_2)$. Figure 4.20 shows, instead, the matrix $\hat{\omega}_{02}\Omega_2$. We have adopted here the following somewhat unusual terminology:

$$\hat{\omega}_{ij} = \omega_{ij}*(\text{the first } \Omega_k \text{ multiplier met in moving} \atop \text{against the arrows in the upper half of Fig. 4.20}). \tag{4.172}$$

Thus, $\hat{\omega}_{ij}$ is only defined with respect to a specific diagram. Furthermore, the same symbol may have a different numerical value at a different location in the diagram. For example, we have just seen that on the extreme left, $\hat{\omega}_{02} = \omega_{02}\omega_{03}$. This symbol appears in two other places to the right. The first one equals $\omega_{02}\omega_{13}$, the second equals $\omega_{02}\omega_{23}$.

Returning now to the main line of the argument, each one of the 5-dimensional β vectors should now be input to a modified DFT tableau of order 5. This time, the β "vectors" generated in phase 1 are of dimensionality 1, namely, scalars. They are to be multiplied by $\hat{\omega}_{i5}\Omega_3 = \hat{\omega}_{i5}$ (Ω_3 is the scalar 1). At this point, the multiplications are actually carried out and the numerical values of the multipliers are needed. Their determination is straightforward. Consider for example the multiplier on the extreme left of Fig. 4.20,

$$\hat{\omega}_{05} = \omega_{05}\hat{\omega}_{02} = \omega_{05}\omega_{02}\omega_{03}.$$

Similarly, for the multiplier on the extreme right

$$\hat{\omega}_{55} = \omega_{55}\hat{\omega}_{12} = \omega_{55}\omega_{12}\omega_{23}$$

and in general, the value of the multiplier at a specific node is the product of all the ω_{ij} terms (stripped of their circumflex) which one encounters in moving from that node up the (inverted) tree structure to the stem.

The multiplications are indicated by the half-circle, half-square shapes stringed along the center line of Fig. 4.20. (These can be regarded as representing the combined three phases of the DFT algorithm of order 1 with $\Omega = \hat{\omega}_{i5}$).

The 36 terms we get after performing the multiplications comprise 6 independent groups resulting from the 6 separate applications of the modified tableau of order 5. The terms of each of these groups are now combined as prescribed in phase 3 of the 5th-order tableau to yield six 5-dimensional vectors. Each of these is actually a term of the form $\hat{\omega}_{i2}\Omega_2\beta_i$ required to complete the computations in the three applications of the second-order tableau. These computations now yield, as tableau outputs, three 10-dimensional vectors which are identical with $\omega_{i3}\Omega_1\beta_i$ of the third-order tableau. Combining these as prescribed by phase 3 of this tableau yields the 30-dimensional Q vector which is just a scrambled version of the desired vector $\hat{F}$.

Figure 4.20, though directly applicable to example (4.170) only, is characteristic of all N values. For larger N, there might be one more level of branching in each half of the diagram and the number of branches per node may be higher. Otherwise, the structure is identical with that of Fig. 4.20.

4.8 Speed Analysis

In Sects. 4.4–6 we constructed the basic DFT tableaus. In Sect. 4.7 we showed how to use them in an efficient algorithm for certain N values. Our purpose in this section is to determine just how fast the resulting algorithm is and present a summary (Table 4.6) of the pertinent parameters for all orders realizable with the constructed tableaus. The basis for these developments is Table 4.5 which

Table 4.5. Summary of basic DFT tableaus

Tableau order	Total number of multiplications	Number of multiplications by 1 or i	Number of additions	Tableau figure	Tableau page
N	M	m	A		
2	2	2	2	4.9	116
3	3	1	6	4.5	112
4	4	4	8	4.10	118
5	6	1	17	4.6	114
7	9	1	36	4.8	116
8	8	6	26	4.14	129
9	11	1	44	4.13	123
16	18	8	74	4.17	136

presents a summary of the tableau parameters. These have been collected from the tableau designations and have been fully explained earlier.

Consider now the algorithm of order $N = \prod_{k=1}^{K} N_k$ as applied to complex data[15]. For each N_k, we read off from Table 4.5 the corresponding number of complex multiplications M_k and complex additions A_k. We are interested in two functions of these variables, namely, the total number of real multiplications $\mathscr{M}$ and the total number of real additions $\mathscr{A}$ for the overall algorithm realized in the order implied in (4.142) (phase 1 of the N_1 tableau realized first; phase 1 of the N_K tableau realized last). With this goal in mind, we turn now to a mathematical formulation of some of the characteristics of the algorithm structure which are quite evident in Fig. 4.20.

We note that the output of phase 1 of the N_1 tableau is a set of M_1 vectors (β_i) of dimensionality v_1. Each of these now generates (at the output of phase 1 of the N_2 tableau) M_2 vectors of dimensionality v_2, and so on. It is obvious therefore that

$$\left.\begin{array}{l} \text{The total output of phase 1 of all the tableaus of order } N_k \\[6pt] \text{consists of } \left(\prod_{i=1}^{k} M_i\right) \text{ vectors of dimensionality } v_k \text{ (4.144).} \end{array}\right\} \quad (4.173)$$

As each one of these vectors is fed to phase 1 of an N_{k+1} tableau, this is also the number of tableaus of order N_{k+1} or, equivalently,

$$\text{The number of tableaus of order } N_k \text{ is } \prod_{i=1}^{k-1} M_i . \quad (4.174)$$

---

15 The computation of the number of arithmetic operations for real data is more involved and will only be briefly discussed later on.

The simplest application of these results is the determination of the number of multiplications. Let $\mathscr{M}_\kappa$ be the total number of (complex) scalar multiplications in phase 2 of the algorithm realized as a cascade of κ stages. The variables which are multiplied are the 1-dimensional β_i's generated in the last stage of the cascade (N_κ). From (4.173) we know that there are $\prod_{i=1}^{\kappa} M_i$ such terms. Hence

$$\mathscr{M}_\kappa = \prod_{i=1}^{\kappa} M_i . \tag{4.175}$$

To get the total number of real multiplications in the overall cascade ($\mathscr{M}$), we note that in each of the counted multiplications, only one of the two factors is complex, the other being real or imaginary (see Table 4.4). Therefore,

$$\mathscr{M} = 2\mathscr{M}_\kappa . \tag{4.176}$$

We have seen in Sect. 4.7 that each of the overall multipliers for the cascade is a product of κ tableau multipliers, one from each tableau of the cascade. Since each tableau has, at least, one multiplier whose value is 1 or i ($m \geq 1$ in Table 4.5), some of the $\mathscr{M}_\kappa$ overall multipliers will be 1 or i. A consideration of the structure of Fig. 4.20 shows that the number of such multipliers is[16]

$$P_\kappa = \prod_{i=1}^{\kappa} m_i . \tag{4.177}$$

Therefore, if one accepts the somewhat more complex programming involved in the special handling of these P_κ trivial multipliers, the DFT can actually be computed with the smaller number of real multiplications

$$\hat{\mathscr{M}} = 2(\mathscr{M}_\kappa - P_\kappa) . \tag{4.178}$$

The summary in Table 4.6 covers both cases [(4.176, 178)][17].

We turn now to the additions count. Let $\mathscr{A}_\kappa$ be the total number of (complex) scalar additions in a cascade of κ stages. Hence, the corresponding number of real additions is

$$\mathscr{A} = 2\mathscr{A}_\kappa . \tag{4.179}$$

---

16 All permissible N values are expressible as $N = H2^r$ (H odd; $0 \leq r \leq 4$). Using this with the values of m_i listed in Table 4.5 yields $P_\kappa = 2r + \delta_{0,r}$.

17 One could argue that in a binary machine, multiplication by $\frac{1}{2}$ is also trivial and should be excluded from the multiplication count. If this attitude is adopted, then the 9th-order tableau will have 3 trivial multiplications ($m = 3$). The effect of this will be quite pronounced for $N = 144$, reducing $\hat{\mathscr{M}}$ from 380 (Table 4.6) to 348.

The simplest way to determine $\mathscr{A}_\kappa$ is through a recursive argument. Assume that we have the result for a cascade of length $(\kappa-1)$ and we add to it one more stage at position κ. This has two effects. First, the additions in the first $(\kappa-1)$ stages will now involve vectors whose dimensionalities are N_κ times their previous values. Hence, the previous $\mathscr{A}_{\kappa-1}$ additions are transformed into $N_\kappa \mathscr{A}_{\kappa-1}$ additions. To this we should add the additions of the last stage. From (4.174), the number of tableaus in the last stage is $\prod_{i=1}^{\kappa-1} M_i = \mathscr{M}_{\kappa-1}$ (4.175). Each such tableau requires A_κ additions of scalars. Hence, the number of scalar additions in the last stage is $\mathscr{M}_{\kappa-1}A_\kappa$ and the total is

$$\left.\begin{aligned}
\mathscr{A}_\kappa &= \mathscr{A}_{\kappa-1}N_\kappa + \mathscr{M}_{\kappa-1}A_\kappa \\
\mathscr{M}_\kappa &= \mathscr{M}_{\kappa-1}M_\kappa \\
\mathscr{A}_1 &= A_1 ; \quad \mathscr{M}_1 = M_1 .
\end{aligned}\right\} \tag{4.180}$$

We have added here a recursive rephrasing of (4.176) as well as the initial conditions to provide a complete prescription for the simultaneous computation of both $\mathscr{A}_\kappa$ and $\mathscr{M}_\kappa$. Equation (4.180) also yields explicit formulae for $\mathscr{A}_\kappa$. For example

$$\begin{aligned}
\mathscr{A}_4 = {} & A_1N_2N_3N_4 \\
& + M_1A_2N_3N_4 \\
& + M_1M_2A_3N_4 \\
& + M_1M_2M_3A_4 .
\end{aligned} \tag{4.181}$$

An important fact clearly indicated by (4.181) is that $\mathscr{A}_\kappa$, unlike $\mathscr{M}_\kappa$, is also a function of the order of the N_k's comprising N. Thus, it is important to find the cascade order minimizing $\mathscr{A}_\kappa$.

With (4.175–180) and Table 4.5 at our disposal, we can now compute $\mathscr{A}, \mathscr{M}$ for any N satisfying (4.142). Table 4.6 presents the results of such computations implemented by a simple computer program which also provides information for the selection of the most efficient cascade ordering. This N_k sequence appears in the last columns of Table 4.6. We provide here room for two sequences since, in some cases, the same values of $\mathscr{M}, \mathscr{A}$ are obtained with two different N_k sequences. In this case, the coice could be governed by arguments other than efficiency.

Each N_k value appears with a bracketed number to its right. This is the ζ_k of (4.163). Thus, Table 4.6 provides both the sequence of tableaus to be realized and the permutation required in the input square of each tableau [see discussion preceding (4.168)].

Note that adoption of the order prescribed in Table 4.6 is quite important. For example, with $N=240$, Table 4.6 states $\mathscr{A}=5016$ with the prescribed order $3;16;5$. If, instead, we adopt the order $5;16;3$, the number of real additions jumps to 5592 – an increase of 11%.

Table 4.6. Summary of DFT algorithm $\left(\mu = \dfrac{\text{time for one real multiplication}}{\text{time for one real addition}}\right)$

Order of DFT	Multiplications by 1 and i, included in count			Multiplications by 1 and i, excluded from count			Number of real additions	Tableau realization sequences $N_k(\zeta_k)$							
	Speed gain function parameters $G(\mu)=G_\infty \dfrac{\mu+1.5}{\mu+R}$		Number of real multiplications	Speed gain function parameters $\hat{G}(\mu)=\hat{G}_\infty \dfrac{\mu+1.5}{\mu+\hat{R}}$		Number of real multiplications				k				k	
N	G_∞	R	$\mathcal{M}$	$\hat{G}_\infty$	$\hat{R}$	$\hat{\mathcal{M}}$	$\mathcal{A}$	1	2	3	4	1	2	3	4
2	1.000	1.000	4	–	–	0	4	2(1)							
3	1.585	2.000	6	2.377	3.000	4	12	3(1)							
4	2.000	2.000	8	–	–	0	16	4(1)							
5	1.935	2.833	12	2.322	3.400	10	34	5(1)							
6	2.585	3.000	12	3.877	4.500	8	36	3(2)	2(1)			2(1)	3(2)		
7	2.183	4.000	18	2.456	4.500	16	72	7(1)							
8	3.000	3.250	16	12.000	13.000	4	52	8(1)							
9	2.594	4.000	22	2.853	4.400	20	88	9(1)							
10	2.768	3.667	24	3.322	4.400	20	88	2(1)	5(3)						
12	3.585	4.000	24	5.377	6.000	16	96	3(1)	4(3)			4(3)	3(1)		
14	2.961	4.778	36	3.331	5.375	32	172	2(1)	7(4)						
15	3.256	4.500	36	3.447	4.765	34	162	3(2)	5(2)						
16	3.556	4.111	36	6.400	7.400	20	148	16(1)							
18	3.412	4.818	44	3.753	5.300	40	212	2(1)	9(5)						
20	3.602	4.500	48	4.322	5.400	40	216	4(1)	5(4)						
21	3.416	5.556	54	3.548	5.769	52	300	3(1)	7(5)						
24	4.585	5.250	48	6.113	7.000	36	252	3(2)	8(3)			8(3)	3(2)		
28	3.739	5.556	72	4.206	6.250	64	400	4(3)	7(2)						
30	4.089	5.333	72	4.330	5.647	68	384	3(1)	2(1)	5(1)		2(1)	3(1)	5(1)	
35	3.325	6.167	108	3.387	6.283	106	666	7(3)	5(3)						
36	4.230	5.636	88	4.653	6.200	80	496	4(1)	9(7)						
40	4.435	5.542	96	5.069	6.333	84	532	8(5)	5(2)						
42	4.194	6.333	108	4.355	6.577	104	684	3(2)	2(1)	7(6)		2(1)	3(2)	7(6)	
45	3.744	6.167	132	3.802	6.262	130	814	9(2)	5(4)						
48	4.964	5.889	108	5.828	6.913	92	636	3(1)	16(11)						
56	4.517	6.528	144	4.927	7.121	132	940	8(7)	7(1)						
60	4.922	6.167	144	5.212	6.529	136	888	3(2)	4(3)	5(3)		4(3)	3(2)	5(3)	
63	3.804	7.111	198	3.843	7.184	196	1,408	9(4)	7(4)						
70	3.973	6.815	216	4.048	6.943	212	1,472	2(1)	7(5)	5(4)					
72	5.048	6.659	176	5.417	7.146	164	1,172	8(1)	9(8)						
80	4.683	6.259	216	5.058	6.760	200	1,352	16(13)	5(1)						
84	4.972	7.111	216	5.163	7.385	208	1,536	3(1)	4(1)	7(3)		4(1)	3(1)	7(3)	
90	4.426	6.848	264	4.494	6.954	260	1,808	2(1)	9(1)	5(2)					
105	4.352	7.463	324	4.379	7.509	322	2,418	3(2)	7(1)	5(1)					
112	4.706	7.198	324	4.951	7.571	308	2,332	16(7)	7(4)						
120	5.756	7.208	288	6.006	7.522	276	2,076	3(1)	8(7)	5(4)		8(7)	3(1)	5(4)	
126	4.440	7.747	396	4.485	7.827	392	3,068	2(1)	9(2)	7(2)					
140	4.621	7.463	432	4.708	7.604	424	3,224	4(3)	7(6)	5(2)					
144	5.214	7.364	396	5.434	7.674	380	2,916	16(9)	9(4)						
168	5.750	8.083	432	5.914	8.314	420	3,492	3(2)	8(5)	7(5)		8(5)	3(2)	7(5)	
180	5.108	7.530	528	5.187	7.646	520	3,976	4(1)	9(5)	5(1)					
210	5.000	8.111	648	5.031	8.161	644	5,256	3(1)	2(1)	7(4)	5(3)	2(1)	3(1)	7(4)	5(3)
240	5.857	7.741	648	6.005	7.937	632	5,016	3(2)	16(15)	5(2)					
252	5.076	8.384	792	5.128	8.469	784	6,640	4(3)	9(1)	7(1)					
280	5.269	8.273	864	5.343	8.390	852	7,148	8(3)	7(3)	5(1)					
315	4.401	8.759	1,188	4.409	8.774	1,186	10,406	9(8)	7(5)	5(2)					
336	5.802	8.580	972	5.899	8.724	956	8,340	3(1)	16(13)	7(6)					
360	5.790	8.383	1,056	5.856	8.479	1,044	8,852	8(5)	9(7)	5(3)					
420	5.648	8.759	1,296	5.683	8.814	1,288	11,352	3(2)	4(1)	7(2)	5(4)	4(1)	3(2)	7(2)	5(4)
504	5.713	9.179	1,584	5.756	9.249	1,572	14,540	8(7)	9(5)	7(4)					
560	5.260	8.831	1,944	5.303	8.905	1,928	17,168	16(11)	7(5)	5(3)					
630	4.931	9.290	2,376	4.940	9.305	2,372	22,072	2(1)	9(4)	7(6)	5(1)				
720	5.753	8.970	2,376	5.792	9.031	2,360	21,312	16(5)	9(8)	5(4)					
840	6.296	9.569	2,592	6.326	9.614	2,580	24,804	3(1)	8(1)	7(1)	5(2)	8(1)	3(1)	7(1)	5(2)
1008	5.644	9.727	3,564	5.669	9.771	3,548	34,668	16(15)	9(7)	7(2)					
1260	5.462	9.820	4,752	5.471	9.836	4,744	46,664	4(3)	9(2)	7(3)	5(3)				
1680	6.173	9.984	5,832	6.190	10.011	5,816	58,224	3(2)	16(9)	7(4)	5(1)				
2520	5.992	10.483	9,504	6.000	10.496	9,492	99,628	8(3)	9(1)	7(5)	5(4)				
5040	5.798	10.939	21,384	5.802	10.948	21,368	233,928	16(3)	9(5)	7(6)	5(2)				

We turn now to the two remaining columns of Table 4.6, namely, G_∞, R. These are the two parameters mentioned in Sect. 4.1 and are required to determine the speed advantage in a specific system.

Let

$$\mu = \frac{\text{time for one real multiplication}}{\text{time for one real addition}}$$

in the specific system considered. We define the gain G of the present algorithm over the (nominal) Cooley-Tukey algorithm by

$$G = \frac{\mu \mathcal{M}_{\text{CT}} + \mathcal{A}_{\text{CT}}}{\mu \mathcal{M} + \mathcal{A}}, \tag{4.182}$$

where $\mathcal{M}_{\text{CT}}$, $\mathcal{A}_{\text{CT}}$ are the Cooley-Tukey parameters introduced in (4.1). Obviously, G is the ratio of the time required by the Cooley-Tukey algorithm to the time required by Winograd's algorithm. We refer to it as the speed gain though it should be realized that it reflects only the time taken by the arithmetic computations. Thus, it does not reflect the higher complexity of Winograd's algorithm which will tend to slow down its execution on a general-purpose computer. However, when we consider a special-purpose DFT machine, then this speed gain function is quite realistic.

G is a function of μ and the four parameters appearing in (4.182). Equation (4.183) is a more convenient two-parameter formulation [see (4.1)]

$$G(\mu) = G_\infty \frac{\mu + 1.5}{\mu + R}, \tag{4.183}$$

where

$$G_\infty = \frac{\mathcal{M}_{\text{CT}}}{\mathcal{M}} \tag{4.184}$$

$$R = \frac{\mathcal{A}}{\mathcal{M}}. \tag{4.185}$$

G_∞ is the asymptotic speed gain that is approached with large μ. R prescribes a pole of the $G(\mu)$ function (at $\mu = -R$) and thus determines the second asymptote. Adding to these two the zero of $G(\mu)$ (at $\mu = -1.5$) makes it very easy to sketch $G(\mu)$ and get a sufficiently precise estimate of it. Finally, if we replace $\mathcal{M}$ by $\hat{\mathcal{M}}$ in (4.182–185), we get the circumflexed entities $\hat{G}, \hat{G}_\infty, \hat{R}$ appearing in Table 4.6.

We conclude with a few words regarding the real data case. The number of arithmetic operations here is not half the number in the corresponding complex data case. The main reason for that is that as the DFT of a real vector is, in general, complex, some of the intermediate entities will be complex too. For example, the 8th-order DFT tableau prescribes

$$\beta_5 = e_2 + ie_5 \tag{4.186}$$

and thus β_5 is complex even for real data. This means of course that the tableaus realizing $\xi_5 = \Omega \beta_5$ have complex data inputs.

Another factor to consider is the fact that the construction of a complex number, given its two components, does not involve any arithmetic addition in spite of the appearance of the plus symbol. In the example previously cited, both e_2 and e_5 are real when the f_i's are real. Hence, the "addition" appearing in (4.186) is free.

4.9 Concluding Remarks

We have tried to present here an orderly development of Winograd's DFT algorithm, starting with the general concept, continuing with the construction of the necessary building blocks, and culminating in a detailed description of their incorporation into the overall algorithm.

In applying the algorithm, Table 4.6 (p. 154) is the starting point as it lists all the permissible N values with their associated performance parameters. Having chosen a particular N, the next step is to consult Table 4.5 (p. 151) in order to locate the specific tableaus called for in Table 4.6. In the actual implementation, one starts with the scrambling of the input vector and then applies phase 1 of the tableaus (in the order prescribed in Table 4.6) to smaller and smaller segments of the data vector in its various partially transformed states. This culminates in single-component "segments" finally being multiplied by constants in phase 2. From here on, the process is reversed in the application of phase 3 of the tableaus: The scalars appearing at the output of phase 2 are combined into vectors of higher and higher dimensionality, finally culminating in an N-dimensional vector which is just a scrambled version of the transformed input vector.

This chapter contains sufficiently detailed information upon which one could base a direct, straightforward, implementation of the above process in either hardware or software. Though we did not consider here any specific implementation, attention should be called to a certain feature of the tableaus specifically designed into them to facilitate their implementation. We are referring here to "in-place" transformation. Consider, for example, the 7th-order tableau (Fig. 4.8). Note that the input components f_2, f_5 are used to compute c_2, c_5 and noting else. Hence, there is no need to assign additional storage for c_2, c_5. They may be stored back into the f array, overwriting f_2, f_5. The only requirement is for a temporary storage for, say, f_2 so that after we store c_2 we still have f_2 available for the computation of c_5. Note that even if f_2 represents a vector, we still need only a one-word temporary store since the computation is carried out one component at a time.

This property which we have illustrated here with the $(f_2, f_5) \rightarrow (c_2, c_5)$ transformation is common to all variables in the DFT tableaus of orders 2, 3, 5, 7. It is also valid for the remaining tableaus, if we regard η as the output vector. The only deviation from the above pattern is that, in some cases, groups of 3 components (rather than 2) have to be considered. In the above tableau, the computation of $\beta_1, \beta_2, \beta_3$ is such an example. Note however, that, even in this case, a one-word temporary store is sufficient.

It should be pointed out that in those applications in which this "in-place" feature is not utilized, the tableaus of orders 4, 9, 8, 16, may be simplified somewhat by permuting the F_i rows/columns in the output square to yield an unscrambled output vector F. In this case, of course, the vector η may be dispensed with.

We turn now to a brief consideration of the precision disadvantage mentioned in Sect. 4.1. We shall see in the Appendix (last paragraph) that some of the manipulations generating the tableaus have a detrimental effect on precision. A similar situation afflicts the computation of δ_1 in some of the tableaus. Examination of (4.84, 85) reveals that the adopted formulation (4.85) involves addition and subtraction of $\Omega\beta_1$. Hence if $|\Omega\beta_1| \gg |\delta_1|$, we are bound to have problems, namely, loss of significant bits in floating-point arithmetic and tendency to overflow in fixed-point arithmetic. Similar addition-subtraction manipulations are dispersed in various disguises throughout the tableaus' derivations.

The effect of these peculiarities of the tableaus is that to guarantee a certain measure of precision in the transformation, we probably need more bits per word than in the Cooley-Tukey algorithm. We do not analyze this effect here but it should be pointed out that the structure of the algorithm as displayed in Fig. 4.20 makes such an analysis relatively simple.

Finally, we conclude with yet another important aspect of the algorithm brought forth in Fig. 4.20 namely, the suitability of its structure to the application of various schemes of parallel processing and pipelining. Indeed, there is fertile ground here for all sorts of ingenious designs and variations. As the algorithm becomes more widely known, more and more of these will undoubtedly materialize.

Acknowledgements. The work described here was carried out at the Jet Propulsion Laboratory, California Institute of Technology, under NASA Contract No. NAS7-100.

The author also wishes to express his thanks to Dr. *L. D. Baumert* formerly of JPL for his helpful comments regarding some number-theoretic aspects of this work.

Appendix: Polynomial Congruences

The derivations in Sect. 4.3 require numerous evaluations of

$$R(x) = S_n(x) \bmod m(x), \tag{A.1}$$

where

$$S_n(x) = \sum_{k=0}^{n} s_k x^k \tag{A.2}$$

and $m(x)$ is a monic polynomial of degree 1 or 2, whose roots lie on the unit circle. We establish here all the needed results.

1) $m(x) = x - x_0$ $(x_0 = \pm 1)$

$R(x)$ must be of degree 0

$$R(x) = r_0 \tag{A.3}$$

and (A.1) is equivalent in this case to

$$S_n(x) = (x - x_0)Q_{n-1}(x) + r_0, \tag{A.4}$$

where $Q_{n-1}(x)$ is a polynomial of degree $(n-1)$. Substituting $x = x_0$ in (A.4), we find

$$R(x) = r_0 = S_n(x_0). \tag{A.5}$$

Note that with $x_0 = \pm 1$, r_0 is a multiplication-free algebraic sum of the coefficients of $S_n(x)$.

2) $m(x) = x^2 - 2x \cos\theta + 1 = (x - x_0)(x - \bar{x}_0)$

$R(x)$ must be of degree 1

$$R(x) = r_0 + r_1 x \tag{A.6}$$

and (A.1) is equivalent to

$$S_n(x) = (x - x_0)(x - \bar{x}_0)Q_{n-2}(x) + (r_0 + r_1 x). \tag{A.7}$$

Hence

$$S_n(x_0) = r_0 + r_1 x_0 \tag{A.8}$$

$$S_n(\bar{x}_0) = r_0 + r_1 \bar{x}_0. \tag{A.9}$$

Note that

$$x_0 = \cos\theta + i \sin\theta = e^{i\theta}. \tag{A.10}$$

Hence, subtracting (A.9) from (A.8) yields

$$2ir_1 \sin\theta = S_n(x_0) - S_n(\bar{x}_0) = \sum_{k=0}^{n} s_k(e^{ik\theta} - e^{-ik\theta})$$

$$\therefore r_1 = \frac{1}{\sin\theta} \sum_{k=0}^{n} s_k \sin k\theta \tag{A.11}$$

$$r_1 = \sum_{k=1}^{n} s_k U_{k-1}(\cos\theta),$$

where $U_m(x)$ is the Chebyshev polynomial of the second kind.

To get r_0, we multiply (A.8) by $\bar{x}_0$ and (A.9) by x_0 and then subtract, getting

$$2ir_0 \sin\theta = x_0 S_n(\bar{x}_0) - \bar{x}_0 S_n(x_0) = \sum_{k=0}^{n} s_k [e^{-i(k-1)\theta} - e^{i(k-1)\theta}]$$

$$= 2is_0 \sin\theta - 2i \sum_{k=2}^{n} s_k \sin(k-1)\theta \qquad \text{(A.12)}$$

$$\therefore r_0 = s_0 - \sum_{k=2}^{n} s_k U_{k-2}(\cos\theta).$$

The specific $m(x)$ polynomials for which $R(x)$ is required are listed in Table A.1 with the corresponding θ values. Note that for these values of θ, $U_k(\cos\theta)$ takes only the values $0, \pm 1$. Hence, both r_0 and r_1 are multiplication-free algebraic sums of the coefficients of $S_n(x)$. The results for all required degrees of $S_n(x)$ are shown in Table A.1.

A specific application of Table A.1 is the following: Given

$$P(x) \bmod m(x) = p_1 x + p_0 \qquad \text{(A.13)}$$

$$Q(x) \bmod m(x) = q_1 x + q_0, \qquad \text{(A.14)}$$

find

$$G(x) = g_1 x + g_0 = [P(x)Q(x)] \bmod m(x). \qquad \text{(A.15)}$$

$G(x)$ can be expressed in terms of (A.13, 14) as follows:

$$G(x) = \{[P(x) \bmod m(x)][Q(x) \bmod m(x)]\} \bmod m(x)$$

$$= [(p_1 q_1) x^2 + (p_1 q_0 + p_0 q_1) x + p_0 q_0] \bmod m(x). \qquad \text{(A.16)}$$

Identifying the bracketed polynomial with $S_2(x)$ in Table A.1, we get the following results:

$$G(x) = [p_1(q_0 + q_1) + p_0 q_1] x + (p_0 q_0 - p_1 q_1) \qquad [m(x) = x^2 - x + 1] \qquad \text{(A.17)}$$

$$G(x) = (p_1 q_0 + p_0 q_1) x + (p_0 q_0 - p_1 q_1) \qquad [m(x) = x^2 + 1] \qquad \text{(A.18)}$$

$$G(x) = [p_1(q_0 - q_1) + p_0 q_1] x + (p_0 q_0 - p_1 q_1) \qquad [m(x) = x^2 + x + 1]. \qquad \text{(A.19)}$$

Table A.1. $S_n(x) \bmod m(x)$ for $S_n(x) = \sum_{k=0}^{n} s_k x^k$

$m(x)$	θ	$S_2(x) \bmod m(x)$	$S_3(x) \bmod m(x)$	$S_5(x) \bmod m(x)$
$x^2 - x + 1$	60°	$(s_1 + s_2)x + (s_0 - s_2)$	–	$(s_1 + s_2 - s_4 - s_5)x + (s_0 - s_2 - s_3 + s_5)$
$x^2 + 1$	90°	$s_1 x + (s_0 - s_2)$	$(s_1 - s_3)x + (s_0 - s_2)$	–
$x^2 + x + 1$	120°	$(s_1 - s_2)x + (s_0 - s_2)$	$(s_1 - s_2)x + (s_0 - s_2 + s_3)$	$(s_1 - s_2 + s_4 - s_5)x + (s_0 - s_2 + s_3 - s_5)$

Table A.2. $[P(x)Q(x)] \bmod m(x)$ for $P(x) \bmod m(x) = p_1 x + p_0$
$$Q(x) \bmod m(x) = q_1 x + q_0$$

$m(x)$	$[P(x)Q(x)] \bmod m(x)$
$x^2 - x + 1$	$[(p_1 + p_0)(q_1 + q_0) - p_0 q_0]x + (p_0 q_0 - p_1 q_1)$
$x^2 + 1$	$[(p_1 - p_0)q_0 + (q_1 + q_0)p_0]x + (q_1 + q_0)p_0 - (p_1 + p_0)q_1$
$x^2 + x + 1$	$[(p_1 - p_0)(q_0 - q_1) + p_0 q_0]x + (p_0 q_0 - p_1 q_1)$

Starting with p_i, q_i, each of these formulae requires 4 multiplications. However, with the proper rearranging of terms, we can replace one of these multiplications with an extra addition, thus getting faster computation. This is accomplished by adding and subtracting $p_0 q_0$ from the g_1 term. This is sufficient for (A.17, 19). In (A.18) we also modify the g_0 term by adding and subtracting $p_0 q_1$. The results are summarized in Table A.2 and, as we see there, 3 multiplications are now sufficient. The $(x^2 + 1)$ case, however, seems to indicate that the price is higher than previously stated, namely, 3 extra additions. In general this is indeed true. However, we intend to apply this result to a situation where arithmetic operations involving p_0, p_1 only do not count (precomputation). Under these conditions, the price is indeed 1 extra addition.

It should be noted that the higher speed realized by the formulae of Table A.2 is accompanied by the disadvantage of requiring more bits per word. In fixed-point arithmetic we will need more bits to prevent overflow of the intermediate results. In floating-point arithmetic, we will need more bits to prevent loss of precision. Consider the extreme case in which $p_0 q_0 \gg g_1$. Equation (A.17) would not be affected by that but, in floating point, Table A.2 could yield a value for g_1 which would be pure noise.

References

4.1   J.W.Cooley, J.W.Tukey: Math. Comp. **19**, 297–301 (1965)
4.2   R.Yavne: "An Economical Method for Calculating the Discrete Fourier Transform", *AFIPS Conference Proceedings*, Vol. 33, Part 1 (Thompson Book Co., Washington, D.C. 1968) pp. 115–125
4.3   S.Winograd: Proc. Nat. Acad. Sci. USA **73**, No. 4, 1005–1006 (1976)
4.4   S.Winograd: "The Effect of the Field of Constants on the Number of Multiplications", *Proceedings of the 16th Annual Symposium on Foundations of Computer Science* (The Institute of Electrical and Electronics Engineers, New York 1975) pp. 1–2
4.5   C.M.Rader: Proc. IEEE **56**, 1007–1008 (1968)
4.6   M.Abramowitz, I.Stegun: *Handbook of Mathematical Functions* (Dover, New York 1965) p. 864
4.7   D.E.Knuth: *The Art of Computer Programming*, Vol. 2 (Addison-Wesley, Reading, Mass. 1969) Sect. 4.3.2
4.8   T.Nagell: *Introduction to Number Theory* (Wiley, New York 1951)
4.9   W.J.LeVeque: *Topics in Number Theory*, Vol. 1 (Addison-Wesley, Reading, Mass. 1958)
4.10   I.J.Good: J. R. Stat. Soc. B **20**, 361–372 (1958)

5. Median Filtering: Statistical Properties

B. I. Justusson

With 15 Figures

Median filtering is a nonlinear signal processing technique useful for noise suppression. It was suggested as a tool in time series analysis by *Tukey* [5.1] in 1971 and has later on come into use also in picture processing. Median filtering is performed by letting a window move over the points of a picture (sequence) and replacing the value at the window center with the median of the original values within the window. This yields an output picture (sequence) which usually is smoother than the original one.

The classical smoothing procedure is to use a linear low-pass filter and in many cases this is the most appropriate procedure. However, in certain situations median filtering is better and two of its main advantages are: I) Median filtering preserves sharp edges, whereas linear low-pass filtering blurs such edges. II) Median filters are very efficient for smoothing of spiky noise. We illustrate these properties in Fig. 5.1.

The chief objective of this chapter is to present various theoretical results about median filtering. It is hoped that these results are helpful in judging the practical applicability of median filters.

In Sect. 5.1 basic definitions concerning median filters are given. The ability of median filters to reduce noise is examined in Sect. 5.2, and formulas which yield quantitative information about how much the noise is reduced are presented. White noise, nonwhite noise, impulse noise, and salt-and-pepper noise are considered. In Sect. 5.3 we compare the performance of moving averages and median filters on pictures of the form "edge plus noise". Second-order properties of median filters on random noise are treated in Sect. 5.4. Exact results are given for white-noise input, whereas for nonwhite noise, approximate results are obtained through limit theorems. Frequency response is discussed for simple cosine wave input and also for more general input. In

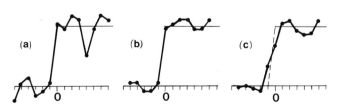

Fig. 5.1. (a) Edge plus noise sequence, (b) after median filtering, (c) after moving average filtering, $n=3$

Sect. 5.5 we present some modifications of median filters which also have the property of preserving edges but differ from simple median filters in other respects. Further uses of medians and other order statistics are discussed in Sect. 5.6.

We conclude this introduction with a short review of earlier work on medians and median filtering.

Medians have long since been used and investigated in statistics as alternatives to sample means in estimation of population means. Most investigations have been concerned with medians and other order statistics of independent random variables; see, e.g., the well-known monographs [5.2, 3]. However, medians of dependent variables have also been treated (see [5.4], where further references are given).

As mentioned above, moving (or running) medians were proposed by *Tukey*, who used them to smooth economic time series. *Tukey* also discussed iterated median filtering and pointed out that median filtering preserves large sudden changes of level (i.e., edges) in time series. *Rabiner* et al. [5.5] and *Jayant* [5.6] used moving medians in speech processing to clean pitches from noise. *Frieden* [5.7] developed a signal processing technique for edge enhancement in which a median filter is used as post-filter to eliminate spurious oscillations. Median filters were later on applied to pictures by several authors. In 1975 *Pratt* examined the effects of median filtering of pictures with normal white noise and with impulse noise. He also investigated the effects of different shapes of the filter windows. His report has been published in [Ref. 5.8, Sect. 12.6]. Median filters were used to correct scanner noise by *Wecksung* and *Campbell* [5.9].

5.1 Definition of Median Filters

5.1.1 One-Dimensional Median Filters

The *median* of n numbers $x_1, ..., x_n$ is, for n odd, the middle number in size. For n even we define it as the mean of the two middle numbers. For n even other definitions can be found in the literature but since they differ only slightly and since n will be odd in most of our applications we will not discuss this topic further. We denote the median by

$$\text{Median}(x_1, ..., x_n). \tag{5.1}$$

For example: Median$(0, 3, 4, 0, 7) = 3$.

A *median filter of size n* on a sequence $\{x_i, i \in \mathbb{Z}\}$ is for n odd defined through

$$y_i = \text{Median}\, x_i \triangleq \underset{n}{\text{Median}}(x_{i-v}, ..., x_i, ..., x_{i+v}), \quad i \in \mathbb{Z}, \tag{5.2}$$

where $v=(n-1)/2$ and $\mathbb{Z}$ denotes the set of all natural numbers. Other terminology in use is *moving medians* and *running medians*.

It is easily seen that this median filter preserves edges, whereas the corresponding moving average filter

$$z_i=(x_{i-v}+ \ldots +x_i+ \ldots +x_{i+v})/n, \quad i\in\mathbb{Z}, \tag{5.3}$$

changes an edge into a ramp with width n (see Chap. 6).

5.1.2 Two-Dimensional Median Filters

Digital pictures will be represented by sets of numbers on a square lattice $\{x_{ij}\}$ where (i, j) runs over $\mathbb{Z}^2$ or some subset of $\mathbb{Z}^2$.

A *two-dimensional median filter with filter window A* on a picture $\{x_{ij}, (i, j)\in\mathbb{Z}^2\}$ is defined through

$$y_{ij}= \underset{A}{\text{Median}}\, x_{ij} \triangleq \text{Median}\,[x_{i+r, j+s}; (r, s)\in A], \quad (i, j)\in\mathbb{Z}^2. \tag{5.4}$$

Various forms of filter windows A can be used, e.g., line segments, squares, circles, crosses, square frames, circle rings. Some examples are shown in Fig. 5.2. The "circle rings" in Fig. 5.2f have been chosen so as to make the number of points in each ring approximately proportional to the area of a corresponding perfect circle ring.

The definitions of median filters given above do not specify how to compute the output close to end points and border points in finite sequences and

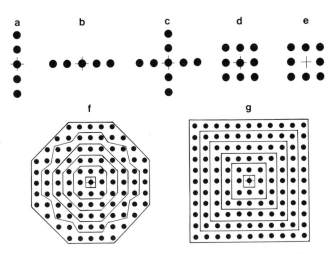

Fig. 5.2a–g. Filter windows. (**a, b**) Line segments, (**c**) a cross, (**d**) a square, (**e**) a square frame, (**f**) circles and circle rings, (**g**) squares and square frames

pictures. One simple convention, which we have used for the generated pictures, is to compute medians of the points lying inside both the picture and the window. Thus for points close to the borders the medians will be computed from fewer than the number of points in A.

5.1.3 Edge Preservation

By an *edge picture* we mean a picture in which all points on one side of a line have a common value a and all points on the other side of the line have a common value b, $b \neq a$. The following result presents a fundamental property of median filters.

If the window set A is symmetric around the origin and includes the origin, i.e., if

$$(r, s) \in A \Rightarrow (-r, -s) \in A, \tag{5.5}$$

$$(0, 0) \in A, \tag{5.6}$$

then the median filter (5.4) preserves any edge picture.

We refer to Chap. 6 by *Tyan* for a thorough discussion of the effects of median filtering of other deterministic signals than just edge pictures. Conditions (5.5, 6) are fulfilled for the windows in Fig. 5.2 except for square frames and circle rings which do not include the origin. However, square frames and circle rings will change edges only slightly. Unless otherwise stated filter windows will be assumed to fulfill conditions (5.5, 6). Note that these conditions imply that the number of points n in A is odd.

5.2 Noise Reduction by Median Filtering

As stated before, median filters can be used for noise suppression. *Pratt* [5.8] discussed in a rather qualitative way their effect on white noise and impulse noise. We shall here present some variance formulas which yield quantitative information about how much the noise will be suppressed.

Median filters are nonlinear and this fact complicates the mathematical analysis of their performance. It is not possible to separate signal effects and noise effects as simply as for linear filters. Throughout this section we confine ourselves to the easiest case with a constant signal.

5.2.1 White Noise

White-noise model The picture values $\{x_{ij}\}$ or the sequence values $\{x_i\}$ are independent identically distributed (i.i.d.) random variables with mean m

$$x = m + z, \tag{5.7}$$

where $E(z)=0$ and thus $E(x)=m$. Let $F(x)$ and $f(x)=F'(x)$ denote the distribution and density functions of the x variables.

Next we write down two well-known results from probability theory about medians of i.i.d. random variables (see [Ref. 5.3, Chaps. 2, 9]).

The density of $y=\text{Median}(x_1, ..., x_n)$ is for n odd given by

$$g(y)=n\binom{n-1}{(n-1)/2}f(y)F(y)^{(n-1)/2}[1-F(y)]^{(n-1)/2}. \tag{5.8}$$

The distribution of $y=\text{Median}(x_1, ..., x_n)$ is for large n approximately normal $N(\tilde{m}, \sigma_n)$ where $\tilde{m}$ is the theoretical median, i.e., is determined by $F(\tilde{m})=0.5$, and

$$\sigma_n^2 = \frac{1}{n4f^2(\tilde{m})} \approx \text{Var}[\text{Median}(x_1, ..., x_n)]. \tag{5.9}$$

For small n values one often gets a better approximation of the variance by replacing the factor $1/n$ in (5.9) with $1/(n+b)$, where $b=1/[4f^2(\tilde{m})\sigma_x^2]-1$. This modification is obtained by chosing b so that (5.9) becomes exact for $n=1$.

The above results yield information for one-dimensional as well as two-dimensional filtering by taking n equal to the number of points in the filter window. It may be noted that if $f(x)$ is symmetric around m then (5.8) will also be symmetric around m and thus the following simple formula holds:

$$E[\text{Median}(x_1, ..., x_n)]=E(x_i)=m. \tag{5.10}$$

Example 5.1: Uniform Distribution. If the x variables are i.i.d. and uniformly distributed on $[0, 1]$ then one can compute the variance of the median exactly using (5.8).

$$\text{Var}[\text{Median}(x_1, ..., x_n)]=\frac{1}{(n+2)\cdot 4}=\frac{\sigma_x^2}{n+2}\cdot 3. \tag{5.11}$$

Formula (5.9) with small sample modification yields the same result.

Example 5.2: Normal Distribution. If the x variables are i.i.d. and $N(m, \sigma)$ then $\tilde{m}=m$ and the variance can only be computed by numerical integration using (5.8). The entries $(n, m, k)=(n, n, n)$ in Table 5.4 give variances for medians of $N(0, 1)$ variables. Formula (5.9) together with the modification for small n yields

$$\text{Var}[\text{Median}(x_1, ..., x_n)]\approx\frac{\sigma^2}{n+\pi/2-1}\cdot\frac{\pi}{2}, \quad n=1, 3, 5, ... \tag{5.12}$$

This formula has good accuracy for all odd n.

The average (mean) $\bar{x}$ of n i.i.d. random variables has variance σ^2/n. Equation (5.12) thus yields that for normal white noise the variance of the median is approximately $(\pi/2-1)=57\%$ larger than that of the mean. Hence a moving average suppresses normal white noise somewhat more than a median filter with the same filter window. Otherwise formulated: To get a median filter with the same variance as a given moving average one has to take 57% more points in the filter window.

Figure 5.3 illustrates median filtering and moving average filtering with square 3×3 windows. Each picture has 45×30 points and each point is 1×1 mm. (a_1) is the original test picture. (b_1, c_1, d_1) have been obtained by changing the grey-scale of (a_1) and adding normal white noise with σ-values $h/5, h/3, h$ where h is the largest edge height. Figure 5.3 will be discussed further in Sect. 5.3.

Example 5.3: Double Exponential Distribution. Let the x variables have a double exponential distribution with mean m and variance σ^2, i.e., have density function

$$f(x) = \frac{\sqrt{2}}{\sigma} e^{-\sqrt{2}\cdot|x-m|/\sigma}, \qquad x \in R. \tag{5.13}$$

Then by (5.9) the asymptotic variance of Median $(x_1, ..., x_n)$ is

$$\sigma_n^2 = \frac{\sigma^2}{(n-1/2)} \cdot \frac{1}{2} \approx \mathrm{Var}\,(\mathrm{Median}) \tag{5.14}$$

which is 50% smaller than the variance σ^2/n of the mean $\bar{x}$. Thus, for this type of noise the median is a better estimator of m than the mean $\bar{x}$. In fact the median is the maximum-likelihood estimator of m and hence the mean-square-error optimal estimator of m (at least asymptotically). The fact that the median here is the maximum-likelihood estimator is an easy consequence of the general result that the median is the least-absolute-deviation estimator of the center of a distribution, i.e.,

$$\min_{a} \sum_{i=1}^{n} |x_i - a| \tag{5.15}$$

is attained for $a = $ Median $(x_1, ..., x_n)$. The above results suggest the more general conclusion that medians are better than means for suppression of noise with heavy-tailed distributions. One extreme is impulse noise which will be discussed in Sect. 5.2.3.

Fig. 5.3a–d. Filtering of pictures with added normal noise $N(0, \sigma)$. (a_1-d_1) Input pictures with $\sigma = 0, h/5, h/3, h$; (a_2-d_2) median filtered pictures; (a_3-d_3) moving average filtered pictures

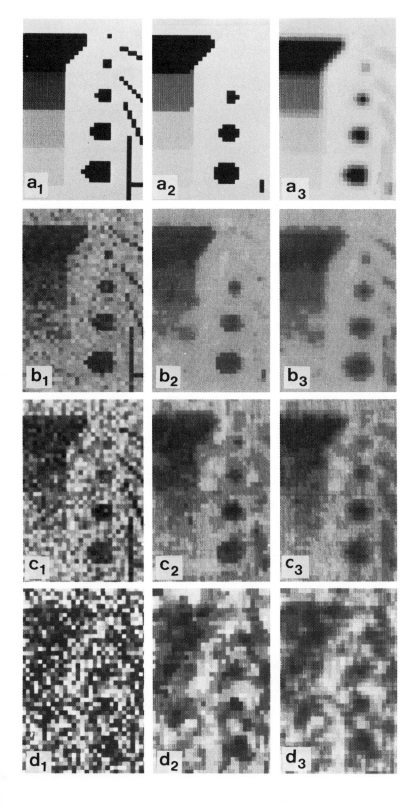

5.2.2 Nonwhite Noise

For input sequences (pictures) that are general random processes (random fields), i.e., with nonindependent variables, one cannot obtain simple exact formulas for the distribution of medians. However, there are limit theorems analogous to (5.9) (see [5.4, 10] where further references can also be found). The conditions needed for the limit theorems are that the processes $\{x_i\}$, $\{x_{ij}\}$ are stationary and mixing. The essence of the mixing condition is that it requires process variables lying far apart to be almost independent (for details see [5.4, 10]). For a stationary mixing normal process with covariance function

$$\text{Cov}(x_i, x_{i+\tau}) = r_x(\tau) = \sigma_x^2 \cdot \varrho_x(\tau), \quad \tau = 0, \pm 1, \dots \tag{5.16}$$

we have the following approximate expression for the variance of a median

$$\text{Var}[\text{Median}(x_1, \dots, x_n)] \approx \frac{\sigma_x^2}{n + \pi/2 - 1} \sum_{j=-(n-1)}^{n-1} \left(1 - \frac{|j|}{n}\right) \arcsin[\varrho_x(j)]. \tag{5.17}$$

For two-dimensional filtering a similar result holds. In Sect. 5.4 these kinds of approximations and limit theorems will be discussed further.

It is interesting to compare (5.17) with the variance of a mean $\bar{x} = (\sum x_i)/n$ of n variables

$$\text{Var}(\bar{x}) = \frac{\sigma_x^2}{n} \sum_{j=-(n-1)}^{n-1} \left(1 - \frac{|j|}{n}\right) \varrho_x(j). \tag{5.18}$$

The similarity of (5.17) and (5.18) is striking. For normal processes with nonnegative correlations

$$\varrho_x(\tau) \geq 0, \quad \tau = 0, \pm 1, \dots \tag{5.19}$$

we obtain, using (5.17, 18) and the fact that $\varrho_x(\tau) \leq \arcsin \varrho_x(\tau) \leq \varrho_x(\tau) \cdot \pi/2$, that for large n

$$1 \leq \frac{\text{Var}(\text{Median})}{\text{Var}(\bar{x})} \leq \frac{\pi}{2}. \tag{5.20}$$

(This result holds for two-dimensional filtering as well.) Thus, for normal processes with nonnegative correlations, the variance of a median is at most 57% larger than the variance of a mean. For processes with both negative and positive correlations the value of the variance quotient in (5.20) can, however, be much larger than $\pi/2$. This point is illustrated in Table 5.1 which presents values of the variance quotient for a normal AR(1) process (first-order

Table 5.1. Variance quotients for normal AR(1) processes

a	0.9	0.5	0.0	-0.5	-0.9
$\dfrac{\text{Var (Median)}}{\text{Var }(\bar{x})}$	1.10	1.21	1.57	2.59	6.59

autoregressive process) with

$$\varrho_x(\tau) = a^{|\tau|}, \quad \tau = 0, \pm 1, \dots . \tag{5.21}$$

Justusson [5.10] reported on findings from computer simulations of normal AR(1) processes which show that also for small n the quotient values in Table 5.1 are valid approximately, except for $a = -0.9$. For $a = -0.9$, $n = 9$ a quotient value as large as 14.9 was obtained.

5.2.3 Impulse Noise and Salt-and-Pepper Noise

By *impulse noise* we mean that a signal is disturbed by impulses (spikes), i.e., very large positive or negative values of short duration. Moving medians are well suited to suppress such noise [5.5, 8] provided that the size of the window is chosen to be at least twice the width of the impulses. Then noise impulses which are sufficiently separated will be completely deleted by the median filter. However, impulses lying close to each other may remain.

In picture processing impulse noise stems, for example, from decoding errors, giving rise to black-and-white spots in the picture, and is therefore often called *salt-and-pepper noise*. The errors become especially prominent in very dark or very bright parts of the picture. For such parts of the picture we can derive some simple formulas for the probability of correct reconstruction. We shall consider two models; in the first model all errors get the same value. In the second model the errors get values which are taken at random from the grey scale.

Impulse Noise: Model 1. At each picture point (i, j) an error occurs with probability p independent both of the errors at other picture points and of all the original picture values. An erroneous point gets the (fixed) value d (e.g., the grey scale value for black). Let $\{x_{ij}\}$ be the distorted picture. Then

$$x_{ij} = \begin{cases} d & \text{with probability } p \\ s_{ij} & \text{with probability } 1-p, \end{cases} \tag{5.22}$$

where s_{ij} denotes the values of the original picture.

Assume now that the point (i', j') is located in a part of the picture where the grey values $\{s_{ij}\}$ of the original picture are constant [at least within the window

Table 5.2. Probability of erroneous reconstruction of impulse noise, $1 - Q(n, p)$

Error rate p	Window size n				
	3	5	9	25	49
0.01	0.00030	0.0000099	0.0000000	0	0
0.05	0.00725	0.00116	0.000033	0.0000000	0
0.1	0.028	0.0086	0.00089	0.0000002	0
0.15	0.0608	0.0266	0.00563	0.000017	0
0.2	0.104	0.058	0.0196	0.00037	0.000013
0.3	0.216	0.163	0.099	0.017	0.00165
0.4	0.352	0.317	0.267	0.154	0.0776
0.5	0.500	0.500	0.500	0.500	0.500

A centered at (i', j')], i.e.,

$$s_{i'+r, j'+s} = s_{i'j'} = c \neq d, \quad (r, s) \in A. \tag{5.23}$$

Apply a median filter with window A to $\{x_{ij}\}$

$$y_{ij} = \underset{A}{\text{Median}} (x_{ij}). \tag{5.24}$$

Then the output value $y_{i'j'}$ will be correct, i.e., $y_{i'j'} = s_{i'j'} = c$, if and only if the number of errors within the window A centered at (i', j') is less than half the number of points in A, i.e., less than or equal to $(n-1)/2$, where n is the size of A. The fact that the number of erroneous points in the window has a binomial distribution yields the following result:

$$P\,[\text{correct reconstruction at } (i', j')] = P(y_{i'j'} = s_{i'j'})$$

$$= \sum_{k=0}^{(n-1)/2} \binom{n}{k} p^k (1-p)^{n-k} \triangleq Q(n, p). \tag{5.25}$$

Values of $1 - Q(n, p)$ for some different values of n and p are shown in Table 5.2. It is seen that if the error rate p is not too large, say not larger than 0.3, then median filtering with a rather small window will reduce the portion of errors considerably. Larger windows will of course reduce the noise even more but will also give more signal distortion. The result of a performed median filtering is shown in Fig. 5.4a.

Impulse Noise: Model 2. This model differs from Model 1 above only in the respect that the error points get random, instead of fixed, grey values z_{ij}. These are assumed to be independent random variables with uniform distribution on

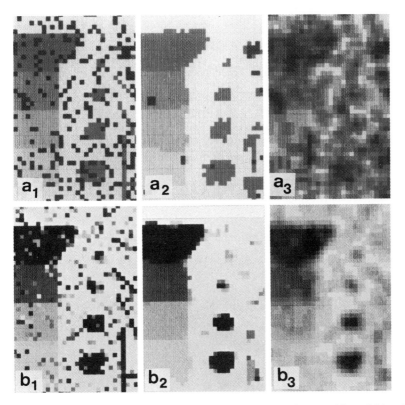

Fig. 5.4a, b. Filtering of impulse noise pictures. (a_1, b_1) Input pictures with model 1 and model 2 noise, respectively, error rates $p = 0.2$; (a_2, b_2) median filtered pictures; (a_3, b_3) moving average filtered pictures

the continuous interval $[0, d]$. In short,

$$x_{ij} = \begin{cases} z_{ij} & \text{with probability } p \\ s_{ij} & \text{with probability } 1 - p. \end{cases} \tag{5.26}$$

To obtain simple formulas we assume that the original picture is totally white (or totally black) in the neighborhood of (i', j'), i.e., $c = 0$ (or $c = d$) in (5.23). This is in a sense the most difficult case for a median filter since all erroneous values then fall on the same side of the correct value. The probability of correct reconstruction is the same as $Q(n, p)$ in (5.25) above, but moreover the magnitude of the remaining errors is diminished. The expected values of the output variables and of the remaining errors are given by

$$E[\text{Median}(x_{i'j'})] = d \cdot \sum_{k=(n+1)/2}^{n} \frac{k - (n-1)/2}{k+1} \binom{n}{k} p^k (1-p)^{n-k} \tag{5.27}$$

and

$$E[\text{Median}(x_{i'j'})|\text{Erroneous reconstruction at } (i', j')]$$

$$= E[\text{Median}(x_{i'j'})]/[1 - Q(n, p)] . \tag{5.28}$$

Proof: Let $N_{i'j'}$ denote the number of errors in A centered at (i', j'). $N_{i'j'}$ is binomially distributed. The conditional distribution of the median at (i', j'), given that $N_{i'j'}$ equals k, is the same as the distribution of $\text{Median}(0, ..., 0, z_1, ..., z_k)$ where the number of zeros is $n - k$, and $z_1, ..., z_k$ are independent and uniformly distributed on $(0, d)$. Further

$$\text{Median}(0, ..., 0, z_1, ..., z_k) = \begin{cases} 0 & \text{if } k \leq (n-1)/2 \\ z_{(r;k)} & \text{if } k \geq (n+1)/2, \end{cases} \tag{5.29}$$

where $z_{(r;k)}$ denotes the rth-order statistic of $z_1, ..., z_k$, and $r = k - (n-1)/2$. Finally

$$E_A[\text{Median}(x_{i'j'})] = \sum_{k=0}^{n} E[\text{Median}(x_{i'j'})|N_{i'j'} = k] P(N_{i'j'} = k)$$

$$= \sum_{k=(n+1)/2}^{n} E\{z_{[k-(n-1)/2;k]}\} \binom{n}{k} p^k (1-p)^{n-k}$$

$$= \sum_{k=(n+1)/2}^{n} d \cdot \frac{k-(n-1)/2}{k+1} \binom{n}{k} p^k (1-p)^{n-k} . \tag{5.30}$$

In the last step we used the expected values of uniform order statistics; see, e.g., [Ref. 5.3, Chap. 3]. Q.E.D.

Figure 5.4b illustrates the case $n = 9$ and $p = 0.2$. According to the above formulas the proportion of erroneous points ought to be reduced from $p = 0.2$ to $1 - Q(n, p) = 0.0196$ and the expected size of an error ought to be reduced from $E(z_{ij}) = 0.5 \cdot d$ to [by (5.28)]

$$\frac{0.00366}{0.0196} \cdot d = 0.187 \cdot d . \tag{5.31}$$

The result of the actual filtering agrees well with those estimates.

Moving averages are not well suited for filtering of impulse noise and salt-and-pepper noise as can be seen in Fig. 5.4. Some filters intended for reduction of salt-and-pepper noise were suggested by *Rosenfeld* and *Kak* [5.11]. We have not carried out a comparison between those filters and the median filter.

A kind of noise which is very similar to impulse noise is *missing line noise* which occurs when pictures are scanned and entire scan lines are lost or when their values are erroneously decoded [5.9]. A median filter with rectangular window with size $n \times m$ (m points on n scan lines) will, as for impulse noise, reduce the number of errors. Assume that erroneous lines occur with probability p independent of other lines, that all values in an erroneous line get the same value d, and that $s_{ij} = c$. Then the probability of correct reconstruction is $Q(n, p)$. Note that here n denotes the number of lines in the rectangular window. Computationally, the simplest window is $m = 1$ but larger m values can sometimes be advantageous for other reasons.

5.3 Edges Plus Noise

So far we have seen that median filters preserve edges (with no noise added) whereas moving averages blur such edges. We have also seen that for normal white noise (with constant background) moving averages reduce the noise somewhat more than median filters with the same window size. In this section we study filtering of edges with added white noise, i.e., sequences or pictures with variables

$$x = s + z, \tag{5.32}$$

where s denotes deterministic signal values which equal 0 on one side of the edge and equal h on the other side, and z are white-noise variables. In Sect. 5.3.1 we compare the effects of moving medians and moving averages on such sequences (pictures). Section 5.3.2 contains the mathematical derivation of the distribution of order statistics on such sequences. The derivation is included in this presentation since it gives an example of the kind of arguments to use when deriving results on order statistics from independent random variables, for example (5.8) and the results in Sect. 5.4.1.

5.3.1 Comparison of Median Filters and Moving Averages

In this subsection we assume that the noise variables z are normal $N(0, \sigma)$. To begin with we consider one-dimensional filtering and assume that the jump of the edge is at point $i = 1$ (cf. Fig. 5.5). Thus, for $i \leq 0$, x_i is $N(0, \sigma)$ and for $i \geq 1$, x_i is $N(h, \sigma)$.

The density $g(x)$ of an n point median with k variables x_i being $N(h, \sigma)$ and $n - k$ variables x_i being $N(0, \sigma)$ can be obtained from (5.37–39) in the next subsection. Though the formula for $g(x)$ is rather complicated, it is quite simple to compute means and standard deviations by numerical integration. This has been done for $\sigma = 1$ and some values of n and k. The results are shown in Table 5.3. Values for $\sigma \neq 1$ can be obtained by a suitable change of scale, and values for $k \geq (n + 1)/2$ can be obtained by symmetry arguments.

The distribution of a moving average is easily obtained as $N(hk/n, \sigma/\sqrt{n})$ where k is the number of variables with $s = h$ within the actual window.

Table 5.3. Moments of medians on edge-plus-noise variables $z_1 + h, \ldots, z_k + h, z_{k+1}, \ldots, z_n$, where the z_is are i.i.d. $N(0, 1)$

			h					
n	k		0.0	1.0	2.0	3.0	4.0	$\geqq 5.0$
3	1	E (Median)	0.000	0.305	0.486	0.549	0.562	0.564
		σ (Median)	0.670	0.697	0.760	0.806	0.822	0.826
9	3	E (Median)	0.000	0.318	0.540	0.626	0.641	0.642
		σ (Median)	0.408	0.424	0.471	0.513	0.527	0.529
5	1	E (Median)	0.000	0.179	0.270	0.294	0.297	0.297
		σ (Median)	0.536	0.551	0.580	0.596	0.600	0.600
5	2	E (Median)	0.000	0.386	0.676	0.808	0.841	0.846
		σ (Median)	0.536	0.560	0.631	0.705	0.740	0.748
25	5	E (Median)	0.000	0.184	0.286	0.312	0.315	0.315
		σ (Median)	0.248	0.256	0.271	0.280	0.282	0.282
25	10	E (Median)	0.000	0.391	0.719	0.900	0.944	0.948
		σ (Median)	0.248	0.260	0.295	0.346	0.371	0.375

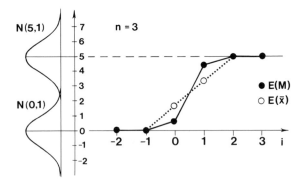

Fig. 5.5. Edge plus noise. Expectations for a moving median (M), and for a moving average ($\bar{x}$), $n = 3$, $h = 5$, $\sigma = 1$

Figure 5.5 shows the successive expected values of medians and means with $n = 3$, around an edge with height $h = 5$. The expected values of the moving average follow a ramp and indicate substantial blurring of the edge. The expected values of the moving median also indicate some blurring, though much smaller than for the moving average.

To be able to compare the efficiencies of filters on edge-plus-noise sequences we need a goodness-of-fit measure. We shall use the average of the mean-square-errors (MSE) at N points close to the edge

$$\frac{1}{N} \sum_i E(y_i - s_i)^2, \tag{5.33}$$

where y_i denotes the filter output values. For the case shown in Fig. 5.5, i.e., $n = 3$ and $N = 2$, the expression (5.33) equals $E(y_0^2)$. In Fig. 5.6 we have plotted

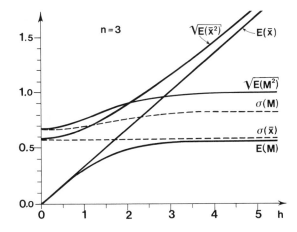

Fig. 5.6. Edge plus noise. Moments for a moving median (M), and for a moving average $(\bar{x})$, $n=3$, $k=1$, $\sigma=1$

$E(y_0^2)^{1/2}$ against the edge height h, both for a median filter and for a moving average filter. It is seen that for $h<2\,(h<2\cdot\sigma)$, the MSE for the moving average is somewhat smaller than the MSE for the median filter, but for $h>3\,(h>3\cdot\sigma)$ the MSE of the median is considerably smaller than that of the mean. This result suggests that moving medians are much better than moving averages for large edge heights, $h>3\sigma$, and that for smaller edge heights there is rather little difference between the two filters. Very similar results are obtained for larger windows, $n=5$, and for two-dimensional filtering, 3×3 and 5×5 windows. The conclusions are also confirmed by the filtered pictures in Fig. 5.3 where h is fixed and $\sigma=0$, $h/5$, $h/3$, and h. The goodness-of-fit measure we have used can only measure the sharpness *across* the edge. It does not say anything about the smoothness of the filtered picture *along* the edge. Figure 5.3 indicates that moving averages give results which are smooth along the edges whereas the median filtered edges are slightly rugged.

We now make some further comments on the behavior of medians as h varies. Throughout we assume that k, the number of variables with mean h, is less than $(n+1)/2$. From Fig. 5.6 and Table 5.3 it is seen that the standard deviations increase with h and have bounded asymptotic values. The expectations are for small h close to the expectations of the corresponding means

$$E(\text{Median}) \approx E(\bar{x}) = h \cdot \frac{k}{n}, \tag{5.34}$$

but for large h they have a bounded asymptote and thus behave radically differently from the expectations of the means. The explanation is that for large h, say $h>4$, the x variables with mean 0 (say $x_1, \ldots, x_{n-k}$ here) will be separated from the x variables with mean h (say $x_{n-k+1}, \ldots, x_n$ here) and then

$$\text{Median}(x_1, \ldots, x_{n-k}, \ldots, x_n) \approx x_{[(n+1)/2, n-k]}, \tag{5.35}$$

where the expression on the right-hand side denotes the $(n+1)/2$th order statistic of $x_1, ..., x_{n-k}$. Expectations and variances of normal order can be found in *Owen* [5.12]. An approximate formula for the expectation of $x_{(r,n)}$ is

$$E[x_{(r,n)}] \approx F^{-1}\left(\frac{r}{n+1}\right), \tag{5.36}$$

where $F(x)$ is the distribution function of the x variables.

Finally, it is of course possible to use the above results in analyzing objects other than edges. The model $x = s + z$ can be used to describe, for example, pulses with added noise.

5.3.2 Distribution of Order Statistics in Samples from Two Distributions

Let $x_1, ..., x_n$ be independent random variables, let $x_1, ..., x_k$ have distribution function $F_1(x)$ and density $f_1(x) = F'_1(x)$, and let $x_{k+1}, ..., x_n$ have distribution function $F_2(x)$ and density $f_2(x) = F'_2(x)$. Then $x_{(r,n)}$, the rth order statistic of $x_1, ..., x_n$ [$r = (n+1)/2$ gives the median] has density

$$g(x) = g_1(x) + g_2(x), \tag{5.37}$$

where

$$g_1(x) = \sum_j k \binom{k-1}{j} \binom{n-k}{r-j-1} f_1(x) F_1(x)^j F_2(x)^{r-j-1}$$
$$\cdot [1 - F_1(x)]^{k-j-1} [1 - F_2(x)]^{n-k-r+j+1}, \tag{5.38}$$

and

$$g_2(x) = \sum_j (n-k) \binom{k}{j} \binom{n-k-1}{r-j-1} f_2(x) F_1(x)^j F_2(x)^{r-j-1}$$
$$\cdot [1 - F_1(x)]^{k-j} [1 - F_2(x)]^{n-k-r+j}. \tag{5.39}$$

The summations are to be carried out over all natural numbers j for which all involved binomial coefficients $\binom{p}{q}$ fulfill $p \geq q \geq 0$.

Proof: The density $g(x)$ of $x_{(r,n)}$ can be obtained by

$$g(x) = \lim_{dx \to 0+} \frac{1}{dx} P[x \leq x_{(r,n)} \leq x + dx]. \tag{5.40}$$

The method of proof is to split up the event in (5.40) into subevents whose probabilities can be calculated. The number of different subevents is calculated using combinatorics. We shall also use the fact that at most one variable x_i can fall into an infinitesimal interval $[x, x+dx]$. The following subevents are used:

A_1: One of $x_1, ..., x_k$ falls into $[x, x+dx]$
A_2: One of $x_{k+1}, ..., x_n$ falls into $[x, x+dx]$
B_j: Exactly j variables of $x_1, ..., x_k$ fall into $(-\infty, x), j=0, ..., k$.

By the law of total probability

$$\frac{1}{dx} P[x \leq x_{(r,n)} \leq x+dx]$$

$$= \sum_j \frac{1}{dx} P[\{x \leq x_{(r,n)} \leq x+dx\} \cap A_1 \cap B_j]$$

$$+ \sum_j \frac{1}{dx} P[\{x \leq x_{(r,n)} \leq x+dx\} \cap A_2 \cap B_j]. \tag{5.41}$$

Consider an event in the first sum. It occurs if and only if one variable of $x_1, ..., x_k$ falls into $[x, x+dx]$, j variables of those remaining of $x_1, ..., x_k$ fall into $(-\infty, x)$, $r-j-1$ variables of $x_{k+1}, ..., x_n$ fall into $(-\infty, x)$, and all remaining variables fall into $(x+dx, +\infty)$. The probability of this event is

$$k f_1(x) dx \cdot \binom{k-1}{j} F_1(x)^j \cdot \binom{n-k}{r-j-1} F_2(x)^{r-j-1}$$

$$\cdot [1 - F_1(x+dx)]^{k-j-1} \cdot [1 - F_2(x+dx)]^{n-k-r+j+1}. \tag{5.42}$$

Inserting this into (5.41) and letting $dx \to 0+$, we obtain $g_1(x)$. By similar arguments one obtains $g_2(x)$ from the second sum in (5.41). In this case one variable from $x_{k+1}, ..., x_n$ falls into $[x, x+dx]$.   Q.E.D.

5.4 Further Properties of Median Filters

In this section we examine covariance and spectral properties of median filters. The obtained results show that the second-order properties of median filters are very similar to those of moving average filters. The last subsection contains some results on sample path behavior.

5.4.1 Covariance Functions; White-Noise Input

Covariance functions for median filtered white-noise sequences were computed in *Justusson* [5.10]. This was done by first deriving formulas for the distribution of pairs of order statistics from partly overlapping samples and then computing covariances by numerical integration using these formulas. The distribution formulas were derived using elaborations of the ideas used in Sect. 5.3.2 but since the formulas are rather complicated we will not reproduce them here. We will, however, present the resulting covariance functions for normal white noise.

Let $\{x_i\}$ be independent $N(0, 1)$-variables and set

$$C(n, m, k) = \text{Cov}[\underset{n}{\text{Median}}\,(x_1, \ldots, x_n), \underset{m}{\text{Median}}\,(x_{n-k+1}, \ldots, x_{n+m-k})], \qquad (5.43)$$

i.e., $C(n, m, k)$ is the covariance of medians from samples of sizes n and m with overlap size k. Numerical values of $C(n, m, k)$ are presented in Table 5.4 together with probabilities $P(n, m, k)$ for the two medians in (5.43) to be equal. For nonoverlapping samples, i.e., $k=0$, we of course have $C(n, m, k)=0$. Note also that $C(n, n, n)= \text{Var}\,(\text{Median})$. Covariances of general normal variables, $N(m_x, \sigma)$, are given by $\sigma^2 \cdot C(n, m, k)$.

The autocovariance function for the output from an n point median filter, $y_i = \text{Median}\,(x_i)$, on normal $N(m_x, \sigma)$ white noise is

$$r_y(\tau) = \text{Cov}(y_{i+\tau}, y_i) = \sigma^2 C[n, n, (n-|\tau|)^+], \qquad (5.44)$$

where $(a)^+ = \max(a, 0)$. Figure 5.7 shows r_y for $n=3$, $\sigma=1$ together with the covariance function r_z of a three-point moving average. The formula for r_z is for general n given by (white-noise input)

$$r_z(\tau) = \sigma^2 \left(1 - \frac{|\tau|}{n}\right)^+. \qquad (5.45)$$

In Fig. 5.7 it is seen that r_y and r_z are similar in form or equivalently that the normalized correlation functions are similar, but that r_y has larger values, i.e., $r_y(0) > r_z(0)$.

Covariance functions for two-dimensional median filters can also be expressed in terms of $C(n, m, k)$. For a $n \times n$ square window we get

$$r_y(\tau_1, \tau_2) = \sigma^2 C[n^2, n^2, (n-|\tau_1|)^+ \cdot (n-|\tau_2|)^+]. \qquad (5.46)$$

Figure 5.8 shows r_y for a 3×3 window. Only the values in the first quadrant are shown. The others can be obtained by symmetry.

The similarity of the correlation functions of moving medians and moving averages is to some extent explained by the relatively high correlation between a median and a mean. (Further explanation is given in the next section.) If $\{x_i\}$ are independent $N(m, \sigma)$ distributed variables then

$$\text{Cov}\left[\text{Median}\,(x_1, \ldots, x_n), \frac{1}{n}\sum_1^n x_i\right] = \text{Var}(\bar{x}) = \frac{\sigma^2}{n}; \qquad (5.47)$$

Table 5.4. Covariances $C(n, m, k)$ and probabilities of equality $P(n, m, k)$ for pairs of medians from $N(0,1)$-samples of sizes n and m with overlap size k [$C(n, n, n) = $ Var (Median)]

n	m	k	C (n, m, k)	P (n, m, k)	n	m	k	C (n, m, k)	P (n, m, k)
1	1	1	1.0000	1.0000	25	25	5	0.0105	0.0253
					25	25	10	0.0214	0.0581
3	1	1	0.3333	0.3333	25	25	15	0.0329	0.1061
3	3	1	0.1177	0.1333	25	25	20	0.0453	0.1972
3	3	2	0.2480	0.3333	25	25	23	0.0538	0.3452
3	3	3	0.4487	1.0000	25	25	24	0.0572	0.4800
					25	25	25	0.0617	1.0000
5	1	1	0.2000	0.2000	49	49	7	0.0040	0.0125
5	3	1	0.0721	0.0857	49	49	14	0.0081	0.0272
5	3	2	0.1494	0.2000	49	49	21	0.0123	0.0456
5	3	3	0.2398	0.4000	49	49	28	0.0166	0.0701
5	5	1	0.0445	0.0571	49	49	35	0.0209	0.1069
5	5	2	0.0914	0.1286	49	49	42	0.0258	0.1797
5	5	3	0.1422	0.2286	49	49	47	0.0296	0.3597
5	5	4	0.2003	0.4000	49	49	48	0.0305	0.4898
5	5	5	0.2868	1.0000	49	49	49	0.0318	1.0000
7	1	1	0.1428	0.1429	81	81	9	0.0018	0.0074
7	3	1	0.0520	0.0635	81	81	18	0.0039	0.0158
7	3	2	0.1069	0.1429	81	81	27	0.0059	0.0256
7	3	3	0.1672	0.2571	81	81	36	0.0079	0.0373
7	5	1	0.0323	0.0433	81	81	45	0.0099	0.0521
7	5	2	0.0660	0.0952	81	81	54	0.0120	0.0721
7	5	3	0.1016	0.1619	81	81	63	0.0142	0.1028
7	5	4	0.1402	0.2571	81	81	72	0.0165	0.1649
7	5	5	0.1848	0.4286	81	81	79	0.0185	0.3658
7	7	1	0.0235	0.0333	81	81	80	0.0188	0.4938
7	7	2	0.0478	0.0722	81	81	81	0.0193	1.0000
7	7	3	0.0732	0.1195	1	1	1	1.0000	1.0000
7	7	4	0.1000	0.1810	9	1	1	0.1111	0.1111
7	7	5	0.1290	0.2698	9	9	9	0.1661	1.0000
7	7	6	0.1621	0.4286	25	1	1	0.0400	0.0400
7	7	7	0.2104	1.000	25	9	9	0.0519	0.1199
9	1	1	0.1111	0.1111	25	25	25	0.0617	1.0000
9	3	1	0.0407	0.0505	49	1	1	0.0206	0.0204
9	3	2	0.0832	0.1111	49	9	9	0.0261	0.0550
9	3	3	0.1286	0.1905	49	25	25	0.0286	0.1157
9	5	1	0.0253	0.0350	49	49	49	0.0318	1.0000
9	5	2	0.0516	0.0758	81	1	1	0.0123	0.0123
9	5	3	0.0791	0.1255	81	9	9	0.0157	0.0320
9	5	4	0.1081	0.1905	81	25	25	0.0171	0.0594
9	5	5	0.1398	0.2857	81	49	49	0.0180	0.1091
9	7	1	0.0186	0.0272	81	81	81	0.0193	1.0000
9	7	2	0.0376	0.0583	121	1	1	0.0083	0.0083
9	7	3	0.0573	0.0949	121	9	9	0.0105	0.0210
9	7	4	0.0778	0.1400	121	25	25	0.0114	0.0372
9	7	5	0.0995	0.1991	121	49	49	0.0119	0.0598
9	7	6	0.1230	0.2857	121	81	81	0.0123	0.1027
9	7	7	0.1499	0.4444	121	121	121	0.0129	1.0000
9	9	1	0.0146	0.0224	2	2	2	0.5000	1.0000
9	9	2	0.0296	0.0476	4	4	4	0.2982	1.0000
9	9	3	0.0449	0.0766	6	6	6	0.2147	1.0000
9	9	4	0.0608	0.1110	8	8	8	0.1682	1.0000
9	9	5	0.0774	0.1536	12	12	12	0.1175	1.0000
9	9	6	0.0949	0.2100	16	16	16	0.0904	1.0000
9	9	7	0.1138	0.2929	20	20	20	0.0734	1.0000
9	9	8	0.1352	0.4444	24	24	24	0.0619	1.0000
9	9	9	0.1661	1.0000	28	28	28	0.0535	1.0000
					32	32	32	0.0471	1.0000
					36	36	36	0.0420	1.0000
					40	40	40	0.0380	1.0000

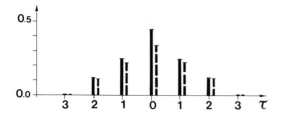

Fig. 5.7. Covariance functions for a three-point moving median (solid lines) and for a three-point moving average (dashed lines) on normal white noise, $N(0, 1)$

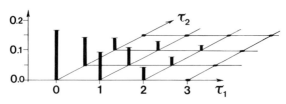

Fig. 5.8. Covariance function for a two-dimensional median filter with a 3×3 square window on normal white noise, $N(0, 1)$

see [Ref. 5.3, p. 31]. Using (5.12) we get for large n the correlation coefficient

$$\varrho(\text{Median}, \bar{x}) \approx \sqrt{\frac{2}{\pi}} = 0.80. \tag{5.48}$$

5.4.2 Covariance Functions; Nonwhite-Noise Input

For autocovariances of median filtered nonwhite noise it is not possible to give general exact formulas. We shall here present some approximate formulas which were derived in *Justusson* [5.10] by considering limit results when the window size tends to infinity. The approximations often work surprisingly well also for small window sizes. For details in the derivations we refer to [5.10].

Assume that $\{x_i\}$ is a stationary mixing sequence with marginal distribution function $F(x)$ with density $f(x)$. We have

$$P[\text{Median}(x_{i-v}, \ldots, x_i, \ldots, x_{i+v}) \leq x] = P\left[\sum_{j=-v}^{v} \text{sign}(x_{i+j} - x) \leq 0\right]. \tag{5.49}$$

By "inversion" of (5.49) one can obtain the following approximate representation formula (Bahadur representation) for large n:

$$y_i = \text{Median}(x_{i-v}, \ldots, x_i, \ldots, x_{i+v})$$

$$\approx \tilde{m} + \frac{1}{2f(\tilde{m})n} \sum_{j=-v}^{v} \text{sign}(x_{i+j} - \tilde{m}), \tag{5.50}$$

where $v = (n-1)/2$ and $F(\tilde{m}) = 0.5$. Thus, a moving average behaves asymptotically as a moving average of sign variables (hard clipped variables). Now an approximation formula for the covariance function of the median filtered sequence can be obtained by computing the covariance function for the moving average to the right in (5.50). This yields

$$r_y(\tau) \approx \frac{1}{nf^2(\tilde{m})} \sum_{j=-(n-1)}^{n-1} \left(1 - \frac{|j|}{n}\right) c_{j+\tau}, \tag{5.51}$$

where $c_k = P(x_0 \leqq \tilde{m}, x_k \leqq \tilde{m}) - 1/4$. For normal noise with covariance function $r_x(\tau) = \sigma^2 \varrho(\tau)$ the quantities c_k can be computed exactly. By doing this and employing the small sample modification mentioned in Sect. 5.2.1 we get

$$r_y(\tau) \approx \frac{\sigma^2}{n + \pi/2 - 1} \sum_{j=-(n-1)}^{n-1} \left(1 - \frac{|j|}{n}\right) \arcsin[\varrho(j+\tau)]. \tag{5.52}$$

In [5.10] we examined the accuracy of the approximation (5.52) for normal white noise and normal AR(1) processes. For input processes with zero, positive, or moderately negative correlations, the accuracy is good even for very small n values. On the other hand, for a process with correlation function $\varrho(\tau) = (-0.9)^{|\tau|}$ the accuracy is bad. Moving medians have only minor smoothing effects on such a process which behaves like $x_i \approx (-1)^i y$ over long time intervals. In fact, an input sequence $x_i = (-1)^i y$, $i \in \mathbb{Z}$, will be unaltered in form by a moving median [though shifted one step for some n values (cf. Chap. 6)]. On the other hand moving averages have a very strong smoothing effect on such a process since the regularly fluctuating x values give rise to cancellations. In general the approximation formulas for the covariances of moving medians can be expected to work well only for sequences on which the median filters act similar to moving averages. In particular they should not be expected to work well on highly oscillatory sequences and on edge sequences.

We can now explain the similarity between the correlation properties of median filters and moving average filters. For large n (5.51) can be approximated by

$$r(\tau) \approx \text{const} \cdot \left(1 - \frac{|\tau|}{n}\right)^+. \tag{5.53}$$

Furthermore, (5.53) holds asymptotically for all moving averages although with different constants. Thus, they have the same normalized correlation function but may have different asymptotic variances.

In Sect. 5.2.2 we mentioned that medians over large windows are approximately normally distributed. This result can be proved by using the Bahadur representation and applying a central limit theorem for stationary mixing processes to the right-hand side of (5.50).

The above ideas can also be applied to two-dimensional median filtering. We get the following Bahadur representation

$$y_{ij} = \underset{A}{\text{Median}}\,(x_{ij}) \approx \tilde{m} + \frac{1}{2f(\tilde{m})n} \sum_{(r,s) \in A} \text{sign}(x_{i+r,j+s} - \tilde{m}), \qquad (5.54)$$

where n is the size of the window A. For normal noise $\{x_{ij}\}$ the corresponding approximation of the covariance function $r_y(\tau_1, \tau_2)$ becomes

$$r_y(\tau_1, \tau_2) \approx \frac{\sigma^2}{n + \pi/2 - 1} \sum_{(r,s) \in A} \sum_{(r',s') \in A} \arcsin[\varrho(\tau_1 + r - r', \tau_2 + s - s')]. \qquad (5.55)$$

For some windows (5.55) can be simplified further.

5.4.3 Frequency Response

Impulse response, step response, and frequency response functions are often used to describe filters. Since a median filter wipes out impulses and preserves edges, the impulse response is zero, and the step response is unity. We shall in this subsection compute the power spectrum distribution of median filtered cosine waves in the cases $n = 3$ and $n = 5$.

To begin with we consider continuous time filtering. Let, for $0 \leq \omega_0 \leq \pi$,

$$x(t) = \cos(\omega_0 t), \quad t \in R, \qquad (5.56)$$

$$y(t) = \text{Median}[x(t-1), x(t), x(t+1)], \quad t \in R. \qquad (5.57)$$

It is easily seen (cf. Fig. 5.9a) that

$$y(t) = \begin{cases} \cos[\omega_0(t-1)] & 0 \leq t \leq 1/2 \\ \cos(\omega_0 t) & 1/2 \leq t \leq T/2 - 1/2 \\ \cos[\omega_0(t+1)] & T/2 - 1/2 \leq t \leq T/2, \end{cases} \qquad (5.58)$$

where $T = 2\pi/\omega_0$. Since $y(t)$ is an even periodic function with period T, (5.58) determines $y(t)$ for all t.

The variance of $y(t)$ is obtained by straightforward integrations

$$\sigma_y^2 = \frac{1}{T} \int_0^T y^2(t)\,dt \qquad (5.59a)$$

$$= \frac{1}{2}\left[1 + \frac{1}{\pi}(\sin 2\omega_0 - 2\sin\omega_0)\right], \quad 0 \leq \omega_0 \leq \pi. \qquad (5.59b)$$

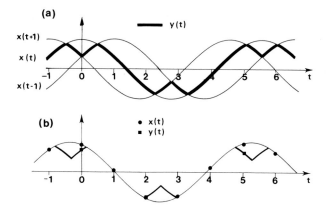

Fig. 5.9a, b. Median filtering of a cosine wave, $n=3$. (a) Continuous time, $x(t)=\cos(\omega_0 t)$; (b) discrete time, $x(t)=\cos(\omega_0 t+\theta)$, $\omega_0=9/8$, period $T=2\pi/\omega_0=5.6$, $y(t)=\text{Median}\,[x(t-1),x(t), x(t+1)]$

We can expand $y(t)$ as a Fourier series with coefficients

$$c_k=\frac{1}{T}\int_0^T e^{itk\omega_0}y(t)dt, \quad k=0, \pm 1, \pm 2, \dots. \tag{5.60}$$

Simple calculations give

$$c_1=\frac{1}{2}\left(1-\frac{\omega_0}{\pi}-\frac{\omega_0}{\pi}\cos\omega_0\right), \quad 0\leq\omega_0\leq\pi. \tag{5.61}$$

Note that $2c_1^2$ is the spectral mass (effect) at the frequencies $\pm\omega_0$, and that σ_y^2 is the total spectral mass (effect). Figure 5.10a shows σ_y^2 and $2c_1^2$ for $0\leq\omega_0\leq\pi$. Also shown is σ_y^2 for a corresponding three-point moving average ($2c_1^2=\sigma_y^2$ for linear filters).

It is seen that for low-frequency cosine input ($\omega_0\leq 2\pi/3$ or $T\geq n$) the moving median and moving average have similar response, whereas for $\omega_0>2\pi/3$ both σ_y^2 and $2c_1^2$ increase for the median, and at $\omega_0=\pi$ they reach the same value as for $\omega_0=0$. This is explained by the fact that a three-point median filter will preserve the form of a sequence $x_i=(-1)^i x$ but shift it one step.

In discrete time the choice of zero as phase angle in (5.56) is rather arbitrary. If, instead, the phase angle θ is chosen at random with uniform distribution on $[0, 2\pi]$, one obtains a stationary process

$$x(t)=\cos(\omega_0 t+\theta), \quad t=0, \pm 1, \pm 2, \dots, \tag{5.62}$$

and the median filtered output (cf. Fig. 5.9b) will have the same covariance properties as the continuous time process (5.57). The power spectrum

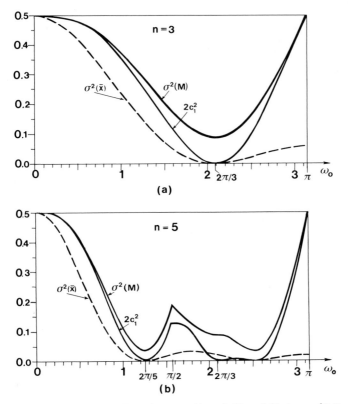

Fig. 5.10a, b. Filtering of a cosine wave. (a) $n=3$, (b) $n=5$. Variance $\sigma^2(M)$ and spectral mass at ω_0, $2c_1^2$, for a median filtered cosine, and variance $\sigma^2(\bar{x})$ for a moving average filtered cosine

distribution of the median filtered sequence is obtained from (5.60) by folding the spectral masses c_k^2 on $k\omega_0$ into the interval $(-\pi, \pi]$. For ω_0 values which are rational multiples of π, this may result in overlays of spectral masses, but apart from this we can still interpret $2c_1^2$ as the spectral mass (effect) on $\pm\omega_0$.

We have carried out the same kind of analysis for median filters with $n=5$. The results are presented in Fig. 5.10b. The derivations were rather cumbersome. The results agree with simulation experiments performed by *Velleman* [5.13]. For frequencies $\omega_0 \leq 2\pi/5$, i.e., $T \geq n=5$, the spectral responses of moving medians and moving averages are similar. For general n this will hold for $T \geq n$. Results for larger values of n can be obtained by numerical integration of (5.59a) and (5.60) with $y(t)$ as in (5.57).

Two-dimensional median filtering with 3×3 and 5×5 square windows of a cosine wave along the axis of the point grid yields the same variances and spectral components as shown above.

Median filters are nonlinear and hence the above frequency response functions for single cosines do not correspond to transfer functions for general

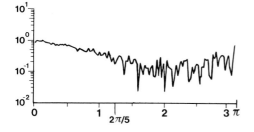

Fig. **5.11.** Cross section of an empirical "transfer function" for a 5×5 median filtered picture (reproduced from [Ref. 5.14, Fig. 3.6] with kind permission of G. Heygster)

signals (mixtures of cosines). *Heygster* [5.14] has computed empirical "transfer functions" as quotients of Fourier transforms of output and input pictures. One example for a 5×5 median filter is shown in Fig. 5.11. It shows a cross section of the bivariate transfer function with one spectral coordinate set to zero. Note that for frequencies $\omega_1 \leq 2\pi/5$ the curve is very smooth, and can thus be interpreted as a transfer function, whereas for $\omega_1 \geq 2\pi/5$ the curve is very irregular due to interferences from different frequencies. Of course these transfer functions depend on the actual input pictures chosen. Heygster's results confirm the previously noted similarity of spectral response of moving medians and moving averages for frequencies $\omega \leq 2\pi/n$ or $T \geq n$.

5.4.4 Sample-Path Properties

We have already discussed the effect of median filtering of edges and highly oscillatory sequences. Excluding such sequences the general shape of median filtered sequences is similar to that of moving average filtered sequences. This follows from the Bahadur representation (5.50).

Although the general appearance of moving median and moving average sequences is the same there are differences in details. Several authors have noted that median filtered sequences include many runs of equal values. The following result gives a simple estimate of the frequency of equal successive values. Let the input sequence $\{x_i\}$ be a stationary mixing random process. Then

$$\lim_{n \to \infty} P[\underset{n}{\text{Median}}(x_i) = \underset{n}{\text{Median}}(x_{i+1})] = 0.5 . \qquad (5.63)$$

Thus, roughly 50 % of all the values in a median filtered sequence are equal to their right neighbors when n is large. (Note that this is not the same as being equal to any one of the two neighboring values.)

Proof: Going from i to $(i+1)$ in the filter computations involves replacing x_{i-v} by x_{i+1+v} in the filter window. If these two variables lie on the same side of $y = \text{Median}(x_i)$ then $\text{Median}(x_{i+1}) = \text{Median}(x_i)$. The mixing condition implies that x_{i-v} and x_{i+1+v} are asymptotically independent. Together with

Median $x_i \approx \tilde{m}$ this gives

$$P[\text{Median}(x_i) = \text{Median}(x_{i+1})]$$
$$\approx P(x_{i-v} < \tilde{m}, x_{i+1+v} < \tilde{m}) + P(x_{i-v} > \tilde{m}, x_{i+1+v} > \tilde{m})$$
$$\approx 0.5 \cdot 0.5 + 0.5 \cdot 0.5 = 0.5. \tag{5.64}$$

which proves (5.63).

For white-noise input the probability in (5.63) can be computed exactly and equals $0.5(1 - 1/n)$, [5.10]. For two-dimensional filtering there is no such simple result like (5.63) but for white-noise input and square windows we can use Table 5.4 to compute probabilities of equality.

$$P[\text{Median}(x_{ij}) = \text{Median}(x_{i,j+1})] = P(n^2, n^2, n^2 - n) \tag{5.65}$$

which for a 3×3 window yields 0.2100 and for a 9×9 window 0.1649.

Another feature of median filtered sequences is the occurrence of small occasional edges, to be compared with moving average filtered sequences which vary in a very "continuous" way. To remove such occasional edges *Rabiner* et al. [5.5] applied a three-point linear filter after the median filtering.

5.5 Some Other Edge-Preserving Filters

In this section we describe some filters which have the same basic properties as median filters: preservation of edges and reduction of noise. All filters to be considered, except those in the last subsection, are variations on the median theme. The previous median filters will here be called *simple median filters*. Variance formulas for the filter output are given for white-noise input.

5.5.1 Linear Combination of Medians

Let A_k, $k = 1, ..., K$ be different windows. Then a *linearly combined median filter* [5.15, 16] is defined through

$$y_{ij} = \sum_{k=1}^{K} a_k \underset{A_k}{\text{Median}}(x_{ij}), \tag{5.66}$$

where a_k are real coefficients. The window sets can, for example, be squares with side lengths $1, 3, ..., 2K - 1$, or circles with diameters $1, 3, ..., 2K - 1$. One can of course also choose window sets which do not include the origin, for example square frames or circle rings (cf. Fig. 5.2).

If all the window sets A_k are symmetric around and include the origin, conditions (5.5, 6), then an edge picture $\{x_{ij}^0\}$ will be preserved in form but the height of the edge will be modified by the factor $\sum a_k$,

$$y_{ij} = \left(\sum_{k=1}^{K} a_k \right) \cdot x_{ij}^0. \tag{5.67}$$

This follows from the fact that each median in the combination preserves the edge. Note that if $\sum a_k = 0$ then $y_{ij} \equiv 0$.

For normal white-noise input $\{x_{ij}\}$ one can compute the variance of y_{ij} using the results in Sect. 5.4.1

$$\text{Var}(y_{ij}) = \sum_{k=1}^{K} \sum_{m=1}^{K} a_k a_m C(n_k, n_m, n_{km}), \tag{5.68}$$

where $C(\cdot, \cdot, \cdot)$ is as in Sect. 5.4.1 and n_k, n_m, n_{km} denote the numbers of points in A_k, A_m, and $A_k \cap A_m$, respectively. Table 5.4 contains values of $C(\cdot, \cdot, \cdot)$ needed when the windows are squares, circle rings, and square frames. The two latter collections of window sets consist of mutually disjoint window sets for which the variance formula simplifies to

$$\text{Var}(y_{ij}) = \sum_{k=1}^{K} a_k^2 C(n_k, n_k, n_k). \tag{5.69}$$

Another way to obtain information about the behavior of linearly combined median filters is to use the fact that simple median filters and moving averages have many similar properties. If one replaces each median in (5.66) with a mean of the same x values, one obtains a linear filter,

$$z_{ij} = \sum_{k=1}^{K} a_k \underset{A_k}{\text{Mean}}(x_{ij}). \tag{5.70}$$

The second-order properties of this filter are easy to calculate and will give at least some information about the properties of the corresponding median filter. One can also go backwards and start with a linear filter with desirable properties and construct a corresponding linearly combined median filter. This is most simply done if the windows are mutually disjoint, e.g., square frames or circle rings.

5.5.2 Weighted-Median Filters

In simple median filters all x values within the window have the same influence on the resulting output. Sometimes one may want to give more weight to the central points. The following type of filter has this possibility. For the sake of simplicity we consider only one-dimensional filters.

The basic idea is to change the number of variables in each window set $(x_{i-v}, \ldots, x_i, \ldots, x_{i+v})$ by taking not one number x_{i+r} but k_r numbers all equal to x_{i+r} and then compute the median of the extended sequence of numbers. We call the resulting filter a *weighted-median filter*.

A simple example may clarify the construction. For $n=3$, $k_{-1}=k_1=2$, $k_0=3$, we have

$$
\begin{aligned}
y_i &= \text{Weighted-Median}(x_{i-1}, x_i, x_{i+1}) \\
&= \text{Median}(x_{i-1}, x_{i-1}, x_i, x_i, x_i, x_{i+1}, x_{i+1}).
\end{aligned}
\tag{5.71}
$$

If the weights are symmetric, $k_{-r}=k_r$, and if k_0 is odd then a weighted-median filter preserves edges.

If the input $\{x_i\}$ is white noise with density $f(x)$ then a weighted median with weights $\{k_r\}$ is for large n approximately normally distributed $N(\tilde{m}, \sigma_n)$ where $\tilde{m}$ is the theoretical median and

$$
\sigma_n^2 = \frac{\sum k_r^2}{(\sum k_r)^2} \cdot \frac{1}{4f^2(\tilde{m})}.
\tag{5.72}
$$

Since this result has not been treated elsewhere we give a brief indication of how it can be proved: By using an extended Bahadur representation [see (5.50)] we get

$$
\begin{aligned}
y_i &= \text{Weighted-Median}(x_i) \\
&\approx \tilde{m} + \frac{1}{2f(\tilde{m})\sum k_r} \sum_{r=-v}^{v} k_r \, \text{sign}(x_{i+r} - \tilde{m}).
\end{aligned}
\tag{5.73}
$$

By the central limit theorem the right-hand side in (5.73) is approximately normally distributed provided that the Lindeberg condition is fulfilled, which, for a weight sequence $k_r = k_r^{(n)}$, here amounts to

$$
\max_r \frac{[k_r^{(n)}]^2}{\sum_i [k_i^{(n)}]^2} \to 0 \quad \text{as } n \to \infty.
\tag{5.74}
$$

This condition essentially requires that none of the factors k_r is considerably larger than the others.

5.5.3 Iterated Medians

Since median filters preserve edges so do iterations of median filters [5.1, 8]. *Tukey* also suggested the following remarkable smoothing procedure: Repeat median filtering until no more changes occur, i.e., until one reaches a median-filtering invariant sequence (fixed-point sequence) (cf. Chap. 6). Note that this invariant sequence need not be constant, in contrast to iterations of moving averages on stationary sequences $\{x_i\}$ which ultimately result in constant sequences.

The statistical properties of iterated medians seem difficult to analyze. We can only present some experiences from simulations of one-dimensional AR(1) processes. For nonnegatively correlated AR(1) sequences, parameter $a \geq 0$, only a few iterations were needed to reach a fixed-point sequence and there were only small changes after the first median filtering. Therefore, it seems likely that the variance formulas for simple medians hold approximately also for iterated medians on processes with nonnegative covariances. For AR(1) processes with alternating positive and negative correlations, $a < 0$, a large number of iterations were needed to reach an invariant sequence and great changes occurred during the filtering process. The final sequences were much smoother and closer to the mean level than the one-step filtered sequences.

When using iterated medians it is of course possible to use different windows in the successive iterations. *Pratt* [5.8] and *Narendra* [5.17] investigated a two-dimensional filtering method in which one first applies a one-dimensional median filter to each line of the picture and then a one-dimensional median filter to each row of the resulting picture, i.e., first

$$z_{ij} = \text{Median}(x_{i,j-v}, \ldots, x_{ij}, \ldots, x_{i,j+v}), \tag{5.75}$$

and then

$$y_{ij} = \text{Median}(z_{i-v,j}, \ldots, z_{ij}, \ldots, z_{i+v,j}). \tag{5.76}$$

The filter is called a *separable median filter*. Its statistical properties can be analyzed theoretically if $\{x_{ij}\}$ is white noise with density $f(x)$ (see [5.17]). The fundamental point is that the z variables in (5.76) are independent since they are computed from x variables in different lines. *Narendra* computed the exact density f_z of the z variables by (5.8) and inserted f_z into the approximate variance formula (5.9) for y_{ij}. We shall present a somewhat simpler formula using the fact that f_z is approximately a normal density $N(\tilde{m}, \sigma_n)$ according to (5.9). Insertion into (5.12) yields

$$\text{Var}(y_{ij}) \approx \frac{1}{n + \pi/2 - 1} \cdot \frac{\pi}{2} \cdot \frac{1}{4 f^2(\tilde{m}) n}. \tag{5.77}$$

For $\{x_{ij}\}$ being normal $N(m, \sigma)$ we get by using (5.77) and the small sample modification in Sect. 5.2.1

$$\text{Var}(y_{ij}) \approx \frac{\sigma^2}{(n + \pi/2 - 1)^2} \left(\frac{\pi}{2}\right)^2. \tag{5.78}$$

For large n this yields output variance $(\sigma/n)^2 \cdot 2.47$ to be compared with the output variance of a simple $n \times n$ median, $(\sigma/n)^2 \cdot 1.57$, and that of a $n \times n$ moving average, $(\sigma/n)^2 \cdot 1$. The variance of a separable median is (roughly) 57% larger than that of an $n \times n$ median and (roughly) 147% larger than that of a moving average.

5.5.4 Residual Smoothing

Assume that the x variables are generated by a signal-plus-noise model

$$x_i = s_i + n_i, \tag{5.79}$$

where the signal $\{s_i\}$ varies slowly compared with the noise $\{n_i\}$. Median filtering gives an estimate of s_i

$$y_i = \text{Median}(x_i) \approx s_i. \tag{5.80}$$

Thus, the residuals

$$\hat{n}_i = x_i - \text{Median}(x_i) \approx n_i \tag{5.81}$$

give estimates of the noise variables. Further median filtering of the residuals could reduce the noise further

$$z_i = \text{Median}(\hat{n}_i) \approx 0. \tag{5.82}$$

Addition of y_i and z_i now gives a hopefully good estimate of s_i,

$$\hat{s}_i = y_i + z_i = \text{Median}(x_i) + \text{Median}[x_i - \text{Median}(x_i)]. \tag{5.83}$$

This *residual smooting* (or *double smoothing*) technique has been suggested and studied by *Rabiner* et al. [5.5], *Velleman* [5.13], and others.

It is easily realized that this smoothing technique preserves edges. Some further combined smoothing methods can be found in [5.18].

5.5.5 Adaptive Edge-Preserving Filters

The filters described earlier are general-purpose smoothing filters. For restricted classes of pictures special filters can be designed. One important class of pictures occurs in remote sensing in connection with classification of picture points into a limited number of classes. The pictures can be described as consisting of compact regions of essentially constant grey level, i.e., underlying the observed picture is an ideal picture consisting of regions of constant grey levels, but the observed pictures have been degraded by addition of noise. Several smoothing algorithms have been proposed for this type of pictures; see, e.g., *Davis* and *Rosenfeld* [5.19] and *Nagao* and *Matsuyama* [5.20] where further references can be found. The main principles used in the proposed algorithms are as follows:

A window is centered at the point (i, j). If there is little variation between the values in the window, it is decided that (i, j) is an interior point and the value at (i, j) is replaced by the average of the values within the window. If, instead, there is much variation between the values in the window, it is decided that (i, j) is a boundary point and (i, j) is assigned the average value of the points which lie on the same side of the border as (i, j) does.

As measures of variation one has used local variances, Laplacians, or gradients. The computational efforts increase very rapidly with window size so it has been suggested to iterate the algorithms instead of using large windows. The algorithms not only preserve sharp edges but also sharpen blurred edges. For example, ramps get changed into step edges. Thus the algorithms do not really "preserve" slanting edges. Some of the algorithms are designed to preserve not only edges but also sharp corners of the region boundaries. For details concerning the aspects touched upon in this subsection we refer to [5.19, 20].

5.6 Use of Medians and Other Order Statistics in Picture Processing Procedures

In the preceding sections we have examined the smoothing ability of median filters. We will now give some examples of how medians can be combined with other picture processing procedures. The examples are taken from edge detection, object extraction, and classification. The final subsection contains a brief account of general order statistics and their use in picture processing.

5.6.1 Edge Detection

The edge-preservation property of median filters makes them suitable as *prefilters* in edge detection, i.e., smoothing the picture with a median filter before applying an edge detector. Many edge detectors are of the type

$$w_{ij} = \begin{cases} 1 & \text{if } |g_1| > \Delta \text{ or } |g_2| > \Delta \\ 0 & \text{otherwise,} \end{cases} \tag{5.84}$$

where g_1, g_2 are gradients at the point (i, j) and Δ is a threshold value. When presmoothing has been performed, it is advisable to use gradients which are symmetric and have span larger than one unit. For example

$$g_1 = y_{i+u,j} - y_{i-u,j} \tag{5.85}$$
$$g_2 = y_{i,j+u} - y_{i,j-u}$$

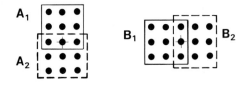

Fig. 5.12. Edge detection. Prefiltering with a median filter (here 3×3) corresponds to a block-type edge detector. The centering point is marked with a small cross

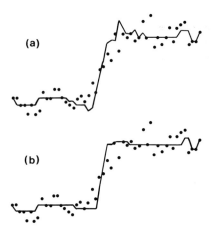

(a)

(b)

Fig. 5.13a, b. Restoration of a step edge. (a) Restoration by a linear filter; (b) restoration by five iterations of a linear filter and a median filter. (Dots show the input data; the output points are connected by lines) (reproduced from [Ref. 5.7, Fig. 2.6] with kind permission of the Optical Society of America © and B. R. Frieden)

with $1/2 \leq u \leq n/2$, $n =$ side length of the smoothing window. Note that using median filtering and such an edge detector is equivalent to using a block-type edge detector in which gradients are computed from medians over blocks,

$$g_1 = \underset{A_1}{\text{Median}}(x_{ij}) - \underset{A_2}{\text{Median}}(x_{ij}),$$

$$g_2 = \underset{B_1}{\text{Median}}(x_{ij}) - \underset{B_2}{\text{Median}}(x_{ij}).$$

$$(5.86)$$

Figure 5.12 illustrates square 3×3 blocks. Gradient span equals 2 ($u = 1$).

Moving medians can also be used as *post-filters* after edge enhancement [5.7]. A sharp edge which has been degraded by a low-pass linear filter can be sharpened by inverse filtering or some modification of inverse filtering. Since the impulse responses of linear filters of inverse type often have large side lobes, they give rise to spurious oscillations at edges (overshots) (cf. Fig. 5.13a). *Frieden* [5.7] used a median filter to eliminate such spurious oscillations, see Fig. 5.13b, in which the linear restoration and median filtering were iterated five times. The window size used, $n = 11$, was equal to the period of the oscillations. Note that the edge in Fig. 5.13b not only has fewer spurious oscillations, but also is sharper than the edge in Fig. 5.13a.

5.6.2 Object Extraction

Extraction of objects in a picture is a common problem in picture processing. We shall consider objects spread out on a smoothly varying background, and

(a)

```
    9 9              9 9
    9 9              9 9



    9                9
```

(b)

```
1  1   1   1   1  1  1  2   3   4   5  6
1  1   1   1   1  1  1  2   3   4   5  6
1  1  [10 10]  1  1  1  2  [12 13]  5  6
1  1  [10 10]  1  1  1  2  [12 13]  5  6
1  1   1   1   1  1  1  2   3   4   5  6
1  1   1   1   1  1  1  2   3   4   5  6
1  1   1   1   1  1  1  2   3   4   5  6
1  1  [10]  1  1  1  1  2  [12]  4   5  6
1  1   1   1   1  1  1  2   3   4   5  6
```

(c)

```
1  1   1   1   1  1  1  2   3   4   5  6
1  1   1   1   1  1  1  2   3   5   5  6
1  1  [1   1]  1  1  1  2  [4   5]  6  6
1  1  [1   1]  1  1  1  2  [4   5]  6  6
1  1   1   1   1  1  1  2   3   5   5  6
1  1   1   1   1  1  1  2   3   4   5  6
1  1   1   1   1  1  1  2   3   4   5  6
1  1  [1]  1   1  1  1  2  [3]  4   5  6
1  1   1   1   1  1  1  2   3   4   5  6
```

(d)

```
1  1   1   1   1  1  1  2   3   4   5  6
1  2   3   3   2  1  1  3   5   6   6  6
1  3  [5   5]  3  1  1  4  [7   8]  7  6
1  3  [5   5]  3  1  1  4  [7   8]  7  6
1  2   3   3   2  1  1  3   5   6   6  6
1  1   1   1   1  1  1  2   3   4   5  6
1  2   2   2   1  1  1  3   4   5   5  6
1  2  [2]  2   1  1  1  3  [4]  5   5  6
1  2   2   2   1  1  1  3   4   5   5  6
```

(e)

```
0  0   0  0   0  0  0  0   0   0   0  0
0  0   0  0   0  0  0  0   0  -1   0  0
0  0  [9  9]  0  0  0  0  [8   8] -1  0
0  0  [9  9]  0  0  0  0  [8   8] -1  0
0  0   0  0   0  0  0  0   0  -1   0  0
0  0   0  0   0  0  0  0   0   0   0  0
0  0   0  0   0  0  0  0   0   0   0  0
0  0  [9]  0  0  0  0  0  [9]  0   0  0
0  0   0  0   0  0  0  0   0   0   0  0
```

(f)

```
0   0    0   0   0  0  0   0    0   0   0  0
0  -1   -2  -2  -1  0  0  -1   -2  -2  -1  0
0  -2  [5   5] -2  0  0  -2  [5    5] -2  0
0  -2  [5   5] -2  0  0  -2  [5    5] -2  0
0  -1   -2  -2  -1  0  0  -1   -2  -2  -1  0
0   0    0   0   0  0  0   0    0   0   0  0
0  -1   -1  -1   0  0  0  -1   -1  -1   0  0
0  -1  [8]-1    0  0  0  -1  [8]-1     0  0
0  -1   -1  -1   0  0  0  -1   -1  -1   0  0
```

Fig. 5.14a–f. Object extraction. (a) Original objects; (b) filter input picture, consisting of objects with added background; (c) 3 × 3 median filtering of the input picture; (d) 3 × 3 moving average filtering of the input picture; (e) subtraction of (c) from the input picture; (f) subtraction of (d) from the input picture

we assume that the objects are relatively well separated. An illustrative example is an aerial photo of stones spread out on hills. An appropriate extraction technique is: first smooth the picture to obtain an estimate of varying background, then subtract this estimate from the original picture. Ideally, this leaves the objects on a constant background (with zero level) and they can then be detected by thresholding. As smoothing operator one can use, for example, a moving median or a moving average, i.e.,

$$y_{ij} = x_{ij} - \underset{A}{\text{Median}}(x_{ij}), \tag{5.87}$$

or

$$z_{ij} = x_{ij} - \underset{A}{\text{Mean}}(x_{ij}). \tag{5.88}$$

The windows can be squares or crosses. If the windows are at least twice as large as the objects then the moving median wipes out the objects and hence the subtraction in (5.87) extracts the objects perfectly. This holds in those parts of the picture where the background is constant. In other parts some distortion of the objects occurs. However, the moving average based filter (5.88) gives considerably more distortion at all points. Figure 5.14 illustrates these aspects. Finally, we note that (5.87) is a linearly combined median filter with $\sum a_k = 1 - 1 = 0$ and thus cancels all edges.

5.6.3 Classification

Multispectral classification of remote sensing pictures is conventionally performed as pointwise classification, i.e., in classifying a point only the spectral values at that point are used. However, for noisy pictures some kind of smoothing is often needed and several methods have been proposed.

Post-smoothing of the result from a pointwise classification through local majority decisions is an easily implemented and frequently used method. Another simple method is presmoothing of each spectral component by moving averages or moving medians. All these methods have the drawback that a pixel which is uniquely classified by its own spectral values can be reclassified due to its neighbors. By combining the spectral values at each point with the median of the neighboring points and using the combined values for classification we obtain an algorithm without the mentioned drawback. Let $x_{ij}^{(k)}$, $k = 1, ..., K$, denote the spectral values at point (i, j) and let

$$y_{ij}^{(k)} = \underset{A}{\text{Median}}[x_{ij}^{(k)}]. \tag{5.89}$$

Then use the extended feature vector

$$(x_{ij}^{(k)}, y_{ij}^{(k)}; k = 1, ..., K)$$

in classification with conventional classification rules for multivariate normal distributions.

The use of medians for smoothing is motivated by its edge preservation which in this context amounts to preservation of class boundaries.

5.6.4 General Order Statistics

The rth-*order statistics* $x_{(r, n)}$ of n numbers $x_1, ..., x_n$ is defined as the rth number in size. In particular $x_{(1, n)} = \min x_i$, $x_{(n, n)} = \max x_i$, and $x_{[(n+1)/2, n]} = \mathrm{Median}(x_i)$. Filtering with *moving order statistics* is performed similar to median filtering: a window is moved over the picture and the rth order statistic of the values within the window is computed. These filters are also called *percentile filters*.

Moving order statistics can be used for shrinking and expanding of objects in pictures (erosion and dilation). Applications have been made, for example, to cell pictures [5.14]. For shrinking $r < (n+1)/2$ is used, e.g., $r = 1$, and for expanding $r > (n+1)/2$, e.g., $r = n$. Note that for $r = (n+1)/2$, i.e., for medians, the average level of the picture is not changed. Shrinking and expanding are often performed by first thresholding the pictures and then applying moving r-out-of-n decision rules. Note that filtering by moving order statistics followed by thresholding gives identically the same results. The intermediate picture may give valuable information. Figure 5.15 illustrates filtering with 3×3 windows and $(r, n) = (2, 9)$.

The statistical literature on order statistics is voluminous. Most of the results on medians presented earlier have counterparts for general order statistics [5.3, 10]. For white-noise input both exact and asymptotic distributions can be obtained, whereas for nonwhite noise only asymptotic results are available. However, the limiting behavior of extreme order statistics [with $r/n \to 0$ or 1, e.g., $x_{(1, n)}$ and $x_{(n, n)}$] are different in nature. In particular, the limiting distributions are nonnormal.

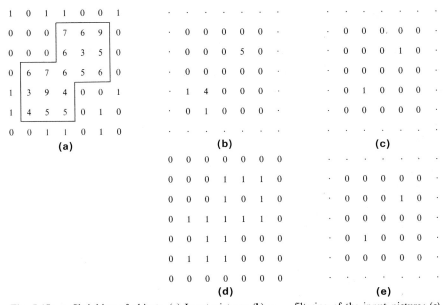

Fig. 5.15a–e. Shrinking of objects. (a) Input picture; (b) $x_{(2, 9)}$ filtering of the input picture; (c) thresholding of (b) with threshold = 4; (d) thresholding of the original picture with threshold = 4; (e) 2-out-of-9 filtering of (d)

For moving order statistics on stationary mixing processes $\{x_i\}$ with r/n staying away from 0 and 1, the mean of the output is asymptotically equal to the λ percentile of the marginal distribution function $F(x)$ of x_i with $\lambda = r/(n+1)$, i.e.,

$$E[x_{(r,n)}] \approx x_\lambda = F^{-1}(\lambda), \qquad (5.90)$$

and the autocovariance function is asymptotically

$$r(\tau) \approx \text{const} \cdot \left(1 - \frac{|\tau|}{n}\right)^+, \qquad (5.91)$$

similar to (5.53) for medians. By (5.90) one can estimate how much the average level of the sequence or picture will be changed. Formula (5.91) shows that moving order statistics have a smoothing effect if r/n is not too close to 0 or 1.

Heygster [5.14] investigated empirical transfer functions (cf. Sect. 5.4.3) also for moving order statistics. For r/n not too close to 0 or 1 these transfer functions have a similar low-pass character as those of moving medians, but for r/n close to 0 or 1 they were not smooth even at low frequencies.

Acknowledgements. The author wishes to thank Prof. B. Rosén and Dr. T. Orhaug for helpful discussions in the course of this work. The author also wants to thank Gunilla Johnson for drawing the figures.

References

5.1   J.W.Tukey: *Exploratory Data Analysis* (Addison-Wesley, Reading, Mass., 1977, preliminary ed. 1971)
5.2   A.E.Sarhan, B.G.Greenberg (eds.): *Contributions to Order Statistics* (Wiley, New York 1962)
5.3   H.A.David: *Order Statistics* (Wiley, New York 1970)
5.4   J.L.Gastwirth, H.Rubin: Ann. Stat. **3**, 1070–1100 (1975)
5.5   L.R.Rabiner, M.R.Sambur, C.E.Schmidt: IEEE Trans. ASSP-**23**, 552–557 (1975)
5.6   N.S.Jayant: IEEE Trans. COM-**24**, 1043–1045 (1976)
5.7   B.R.Frieden: J. Opt. Soc. Am. **66**, 280–283 (1976)
5.8   W.K.Pratt: *Digital Image Processing* (Wiley, New York 1978)
5.9   G.W.Wecksung, K.Campbell: Computer **7**, 63–71 (1974)
5.10 B.Justusson: "Order Statistics on Stationary Random Processes, with Applications to Moving Medians"; Tech. Rpt. TRITA-MAT-1, Royal Institute of Technology, Stockholm, Sweden (1979)
5.11 A.Rosenfeld, A.C.Kak: *Digital Picture Processing* (Academic Press, New York 1976)
5.12 D.B.Owen: *Handbook of Statistical Tables* (Addison-Wesley, Reading, Mass. 1962)
5.13 P.F.Velleman: "Definition and Comparison of Robust Nonlinear Data Smoothing Algorithms"; Tech. Rpt. Economic and Social Statistics Dept., Cornell University (1978)
5.14 G.Heygster: "Untersuchung von Zweidimensionalen Rangordnungsoperatoren im Orts- und Frequenzbereich", in *Bildverarbeitung und Mustererkennung*, ed. by E. Triendl, Informatik Fachberichte, Vol. 17 (Springer, Berlin, Heidelberg, New York 1978) pp. 204–208
5.15 B.Justusson: "Noise Reduction by Median Filtering" in Proc. 4th Intern. Joint Conf. on Pattern Recognition (1978) pp. 502–504
5.16 S.G.Tyan: "Fixed Points of Running Medians"; Tech. Rpt. Dept. of Electrical Engineering and Electrophysics, Polytechnic Institute of New York (1977)
5.17 P.M.Narendra: "A Separable Median Filter for Image Noise Smoothing" in Proc. IEEE Conf. on Pattern Recognition and Image Processing (1978) pp. 137–141
5.18 P.F.Velleman: Proc. Nat. Acad. Sci. USA **74**, 434–436 (1977)
5.19 L.S.Davis, A.Rosenfeld: IEEE Trans. SMC-**8**, 704–710 (1978)
5.20 M.Nagao, T.Matsuyama: "Edge Preserving Smoothing" in Proc. 4th Intern. Joint Conf. on Pattern Recognition (1978) pp. 518–520

6. Median Filtering: Deterministic Properties

S. G. Tyan

With 2 Figures

In this chapter we give some recent results on the properties of median filters. Our approach is a deterministic one. For each median filter, we aim at finding the special class of sequences which are invariant to it. These sequences are called *fixed points* of this particular filter. The purpose may not seem obvious at first; therefore, a few lines on the historic development of median filtering are necessary. Median filtering as it is introduced in [6.1] is used mainly because of the ease of calculating it and its insensitivity to heavy-tailed noise. Then attention is directed to another aspect of median filters, viz., they can preserve sharp transitions or edges in the signals while acting as smoothers [6.2, 3]. This edge-preserving capability makes them attractive smoothers if sharp transitions or edges are frequent and not to be smeared by smoothing. That median filters can preserve edges is tantamount to saying that edges are invariant to median filtering. This is one of the reasons to study fixed points of median filters. If a fixed point has characteristics which are undesirable, then in order that they do not show up at the output it might be necessary to modify the filter such that this particular sequence is not a fixed point or those undesirable characteristics could be suppressed. It can also happen as on certain occasions to be discussed later that an arbitrary input when subject repeatedly to a particular median filter will converge to one of its fixed points. Therefore, by studying fixed points one can get a feel for the effect of a median filter on general input sequences qualitatively if not quantitatively.

Characterization of the fixed points of one-dimensional median filters is presented in Sect. 6.1. It is shown that these fixed points can be divided naturally into two categories. Those of the first category could be considered as some generalized monotonic sequences, while those of the second category are of a more peculiar nature. Then the results obtained for the one-dimensional case will be extended to the two-dimensional domain; however, this part of the theory is still largely incomplete. Finally, an interesting algorithm due to *Huang* et al. [6.4] for doing two-dimensional median filtering is discussed in Sect. 6.4.

6.1 Fixed Points of One-Dimensional Median Filters

We shall denote by MF_{2k+1} the one-dimensional median filter of window length $(2k+1)$, i.e., the input-output relationship is $\{y_n\} = MF_{2k+1}\{x_n\}$, where

$$y_n = \text{median}(x_{n-k}, \ldots, x_n, \ldots, x_{n+k}) \tag{6.1}$$

for all n. Two-sided infinite sequences are assumed. There are relatively few well-known properties of medians which are of direct help to us, and we shall start with two of them:

Property 1. If $x_{-k} \leqq \ldots \leqq x_0 \leqq \ldots \leqq x_k$, then

$$\text{median}(x_{-k}, \ldots, x_0, \ldots, x_k) = x_0.$$

Property 2. If $g(x)$ is monotonic, then

$$\text{median}[g(x_1), \ldots, g(x_{2k+1})] = g[\text{median}(x_1, \ldots, x_{2k+1})].$$

With a median filter as defined in (6.1), one can immediately conclude from Property 1 that a *monotonic sequence* $\{x_n\}$, i.e., $x_n \leqq x_m$ (or $x_n \geqq x_m$) for all $n < m$, is *invariant* to a *median filter* of arbitrary window length. For example, a step function is a monotonic sequence; thus it is invariant to median filters. If a jump is idealized as a step function, then it is easily explained why jumps are preserved under median filtering. Monotonic sequences are only the simplest fixed points of median filters and generalizations of them constitute a more important class of fixed points and will be discussed in this section.

Property 2 implies that scale is irrelevant as far as medians are concerned. In fact, one can reduce the original data sequence $\{x_n\}$ to a bi-valued sequence without losing any information about the medians by considering $\{g_a(x_n)\}$ for all $-\infty < a < \infty$, where

$$g_a(x) = \begin{cases} 1, & x \geqq a \\ 0, & x < a. \end{cases} \tag{6.2}$$

It is obvious that $\{x_n\}$ is a fixed point of a median filter MF_{2k+1} if and only if $\{g_a(x_n)\}$ is a fixed point for all a; however, the input data sequences are much simplified. Property 2 will be most useful when considering two-dimensional data.

As mentioned above, monotonic sequences are fixed points of median filters of arbitrary window lengths; however, the requirement of monotonicity is unnecessarily restrictive. Since a median filter is of fixed and finite window length, intuitively monotonicity is required only for each segment of the window length. In fact, this can be further relaxed.

Definition: A sequence $\{x_n\}$ is *locally monotonic of length m* [LOMO(m)] if $(x_n, \ldots, x_{n+m-1})$ is monotonic for each n.

Obviously, a LOMO(m) sequence is also LOMO(p) provided $p < m$. Suppose $\{x_n\}$ is LOMO(m). Then both $(x_n, \ldots, x_{n+m-1})$ and $(x_{n+1}, \ldots, x_{n+m})$ are monotonic. If $x_n < x_{n+m-1}$ and $x_{n+1} > x_{n+m}$, then we must have $x_i \leqq x_j$ and $x_i \geqq x_j$ for all $n+1 \leqq i \leqq j \leqq n+m-1$, which implies that $x_{n+1} = \ldots = x_{n+m-1}$. Therefore, locally monotonic sequences can be defined alternatively by

Lemma 6.1: If there is any change in trend, then a LOMO(m) sequence must stay constant for at least $(m-1)$ samples.

By using the level function $g_a(x)$ defined previously, one can easily show that a sequence is LOMO(m) if, for all a, the output remains equal to 1 or 0 for at least $m-1$ samples. We have the following theorem on locally monotonic sequences and median filtering.

Theorem 6.1: A LOMO(m) sequence is invariant to MF$_{2k+1}$ for all k, $k \leq m-2$.

Proof: Consider the segment $(x_{n-k}, \ldots, x_{n+k})$, $k = m-2$, which is either monotonic or within it the trend changes with a subsegment of length at least $m-1 = k+1$ staying constant. (See Lemma 6.1.) In the first case, the median of the samples of this segment is obviously equal to x_n. In the second case, the subsegment being of length at least $k+1$ must contain x_n; hence, there are at least $k+1$ elements in the segment equal to x_n and thus the median must be equal to x_n. This completes the proof.

Although LOMO$(k+2)$ sequences are invariant to MF$_{2k+1}$, usually, they do not exhaust the set of fixed points of MF$_{2k+1}$. It has been found that these fixed points belong to two basically different categories. Category I consists of the LOMO$(k+2)$ sequences which have been discussed above. Theorems 6.2 and 6.3 give the characteristics of and the distinctions between the two. The proofs are given in Appendix 6.A.

Theorem 6.2: If $\{x_n\}$ is a fixed point of MF$_{2k+1}$ and if there is a monotonic segment $(x_p, x_{p+1}, \ldots, x_{p+k})$ of length $k+1$, then $\{x_n\}$ is LOMO$(k+2)$.

The above theorem says that if a fixed point of MF$_{2k+1}$ is smooth enough (i.e., monotonic) for a segment of length $k+1$, then it is smooth over the whole length [i.e., LOMO$(k+2)$].

Definition: A sequence is *nowhere* LOMO(k), if it does not contain any monotonic segment of length k.

Theorem 6.3: If $\{x_n\}$ is a fixed point of MF$_{2k+1}$ and if it is nowhere LOMO$(k+1)$, then $\{x_n\}$ is a bi-valued sequence, i.e., x_n can take on only two values.

Theorem 6.3 is the surprising part of median filters and its implication will be discussed further later. For each median filter MF$_{2k+1}$ we shall call those fixed points characterized in Theorem 6.2 Type I fixed points and those in Theorem 6.3 Type II.

Since any segment of two samples is monotonic, by Theorem 6.2 any fixed point of MF$_3$ is LOMO(3). There are no Type II fixed points of MF$_3$. However, Type II fixed points exist for median filters of window length greater than three. In the following, we give a method to generate some of them.

Consider a periodic extension of

$$a_0, a_1, a_2, \ldots, a_k; \quad -a_0, -a_1, -a_2, \ldots, -a_k, \qquad (6.3)$$

where $a_i = 1$ or -1. It is clear that this sequence is a fixed point of MF_{2k+1}. If $(a_0, \ldots, a_k)$ is monotonic, then this sequence is $\mathrm{LOMO}(k+2)$ and hence of Type I. If $(a_0, \ldots, a_k)$ is not monotonic, then this sequence is easily seen to be *nowhere* $\mathrm{LOMO}(k+1)$ and hence is of Type II. Type II fixed points of MF_5 and MF_7 can all be generated by this method, but this is not true for MF_9. For example, the periodic extension of $1, 1, -1, -1$, is a Type II fixed point of MF_9 but it has a periodicity of 4 not of 10 as required by (6.3). It remains unknown whether all Type II fixed points are periodic. Since Type II sequences of MF_{2k+1} are bivalued and tend to fluctuate faster [being nowhere $\mathrm{LOMO}(k+1)$] than Type I sequences [being $\mathrm{LOMO}(k+2)$], they might be considered rather undesirable in a data smoothing problem. In fact, if a segment of a data sequence is highly bivalued and fast fluctuating, then taking the median is of little merit after all. Therefore, it is interesting to consider those generalized median filters which are smoothers built upon simple medians; however, they behave like linear filters, if it is desired, in the presence of Type II fixed points or highly bi-valued fast fluctuating sequences. Before we elaborate on the generalized median filters, we give a simple converse to Theorem 6.1.

Theorem 6.4: If a sequence $\{x_n\}$ is invariant to MF_{2p+1} for all $p = 1, 2, \ldots, k$, then it is $\mathrm{LOMO}(k+2)$.

Proof: By Theorem 6.2 and the fact that any sequence is $\mathrm{LOMO}(2)$, a fixed point of MF_3, i.e., $k = 1$, must be $\mathrm{LOMO}(3)$. Suppose that the theorem holds for $k-1$. Then $\{x_n\}$ which is invariant to MF_{2p+1} for all $p = 1, 2, \ldots, k$, must be $\mathrm{LOMO}(k+1)$; hence, each segment of length $k+1$ is monotonic. By Theorem 6.2, $\{x_n\}$ is $\mathrm{LOMO}(k+2)$.

Rabiner et al. [6.2] observed, for a special case, the possible relation between k and m in Theorem 6.1. *Velleman* [6.5] also described correctly the relation between k and m. He also made the observation that MF_{2k+1} tends to create flat tops (or bottoms) of length $k+1$, which are the very characteristics of $\mathrm{LOMO}(k+2)$ sequences. (See Lemma 6.1.)

So far, we have only treated two-sided infinite sequences; for finite sequences, a few different definitions have been used for the end points [6.6, 7]. For example, with MF_{2k+1}, we may decrease the window size by 2 for each step toward the end once the window center is only k samples away from the end [6.7]. With this definition Theorems 6.1 and 6.2 still hold; however, a slight modification of Theorem 6.3 is necessary for the end points. Since they offer no new insight into the problem, we shall skip it.

Intimate relationship has been observed between median filtering and their fixed points. First, we consider those Type I fixed points. Locally monotonic sequences possess a certain kind of smoothness in terms of monotonicity. Take $\mathrm{LOMO}(m)$ sequences as an example: no change in trend is allowed within any segment of consecutive m samples or equivalently for any change in trend the signal must remain constant for $m-1$ samples, and in between the flats the signal is monotonic. This excludes the possibility of isolated spikes or a burst of them with duration less than or equal to $m-2$. In the words of *Velleman* [6.7], they are not well supported, and median filters of window length $\geq 2m-3$ are capable of removing them. On the other hand, jump-type discontinuities are allowed without regard to the magnitude of the jump so far as the signal is monotonic locally, and hence the next jump which is assumed to be in the opposite direction can not occur within $m-1$ samples. Likewise, median filters of window length $\leq 2m-3$ are capable of preserving it. Of course, not all the properties of LOMO sequences are desirable. For example, the flats which are inevitable for any change in trend can be a problem and median filtering tends to create a great number of them; see [6.6, 8] for examples. Since only local monotonicity is required, signals of this kind do not have to be approximately a low-order polynomial locally. A direct result of this is manifested by median filtering a noise contaminated ramp-like signal. The output may resemble a staircase with irregular steps more than a ramp. Intuitively, by taking the medians, one tends only to restore the monotonicity but not the linearity or other low-order properties intrinsic of the signal [6.7, 9]. In case of this, symmetric linear smoothers of small window length may be advisable after median filtering. Small window is chosen to avoid degrading the edges. As to the flats generated by median filtering, the lost signal can be partially restored by applying reroughing [6.6–8].

We have discussed the advantages and disadvantages of median filtering through the characteristics of locally monotonic sequences which are also the Type I fixed points. Next, we consider the implications of those Type II fixed points. Type II fixed points seem to be of fixed patterns and they rarely show up as intrinsic parts of realistic signals. However, there still is the possibility that portions of a data sequence are fluctuating and bi-valued. If the goal is to recover the weighted mean of each portion, then a median filter is hardly the tool we need. Again, one may use symmetric linear smoothers of small window length after median filtering. However, we are more interested in those nonlinear smoothers which are built upon simple median filters and are free of those Type II fixed points. The main reason is that these nonlinear smoothers are edge preserving. It will be the topic of Sect. 6.2.

6.2 Some Generalized Median Filters

There is another class of sequences associated with median filters which we should be concerned with. These sequences may be called *recurrent points (sequences)*; in general, for a nonlinear smoother T, a sequence is *recurrent* if it is a fixed point of T^m for some $m \geq 2$ but not of T. Here T^m means applying T for m times. For example, the alternating sequence $1, -1, 1, -1, \ldots$ is not a fixed point of MF_3; nevertheless, it is a fixed point of MF_3^2. We know very little about the recurrent sequences of median filters at large. The few examples which we have found suggest that they are very likely to be bi-valued, fluctuating, and of fixed patterns. However, we have no mathematic proof of these observations. If the existence of Type II fixed points makes a median filter useless as a smoother, then one should look for alternatives which are free of them. Similarly, one would like to avoid recurrent points also. The results to this end are scarce, and some of them on MF_3 are given without proof in the following. See Appendix 6.B for details.

Theorem 6.5: Let $a_0 \neq 1$, $a_k \geq 0$, and $\sum_{k=0}^{n} a_k = 1$. Then the smoother

$$T = \sum_{k=0}^{n} a_k MF_3^k \tag{6.4}$$

has only LOMO(3) sequences as its fixed points provided $a_k \neq 0$ for some odd k. If $a_k = 0$ for all odd k, then T has both LOMO(3) and the alternating sequences as its fixed points. For $m \geq 2$, a fixed point of T^m also must be either LOMO(3) or alternating. The alternating sequences cannot be invariant to T^m unless
  1) $a_k = 0$ for all even k and m is even, or
  2) $a_k = 0$ for all odd k.

In view of the above theorem, to avoid the alternating sequence either as a fixed point or as a recurrent point, one has only to consider a smoother T of (6.4) with at least an even k and an odd j such that a_k and a_j are positive. For example, one may use $T_1 = \beta I + (1-\beta)MF_3$, where $0 < \beta < 1$ and I stands for the identity operator, or $T_2 = \beta MF_3 + (1-\beta)MF_3^2 = MF_3 * T_1$ in place of a single MF_3. Indeed, both are free of the Type II fixed points or the recurrent points. Furthermore, we can show that the following is true.

Theorem 6.6: For $T = \beta I + (1-\beta)MF_3$, $0 < \beta < 1$, the sequences $T^m\{x_n\}$ converge pointwise to a LOMO(3) sequence as $m \to \infty$.

The alternating sequences are recurrent points of MF_3 and are also Type II fixed points of MF_5; however, by linearly combining MF_3 and MF_5 one can eliminate them.

Theorem 6.7: Let $T = \beta MF_3 + (1-\beta)MF_5$, $0 < \beta < 1$. Then $\{x_n\}$ is a fixed point iff it is LOMO(4).

It is straightforward to show that smoothers which are constructed by repeated convex combination or compounding [i.e., $\beta T_1 + (1-\beta)T_2$ or $T_1 * T_2$] of median filters of length $\leq 2k+1$ have LOMO($k+2$) sequences as their fixed points. However, it seems to be difficult to decide whether they have any other fixed points, especially, Type II-like fixed points or not.

To understand more about the deterministic properties of median filters or their generalizations, one obviously has to go beyond a mere fixed point theory. *Velleman* [6.5] in his study of the robustness of median filters and related nonlinear filters has taken a different approach to studying the smoothing behavior of these nonlinear filters.

The signals which he has considered are the pure sinusoids. First, a pure zero-phase sinusoid is sampled and then, after median filtering or other nonlinear filtering is done, to compute the power or the amplitude of the fundamental (which has the frequency of the input) and those of its first few harmonics (or their aliases). By this one can study the power transferred by the filter at the input frequency and also that transported to each of its low-order harmonics or aliases. Similar to the power transfer function of a linear system, for a nonlinear filter under study, one can also draw a curve showing the portion of the power transferred at the input frequency with the input being a pure zero-phase sinusoid.

Although the law of superposition does not apply to nonlinear filters, a study of the power transferred and that transported to its harmonics still yields important information about the behavior and merit of each nonlinear filter. For instance, his numerical result shows that the transfer functions of short odd-length median filters have rather large sidelobes. As an example, at a sampling rate of 128 samples per second and with MF_5, large sidelobes appear near frequencies 32 and 64 Hz. Interestingly, this phenomenon is closely related to the fixed points and the recurrent points of MF_5, and we shall discuss it in the following.

Let the input x_n be a sampled sinusoid of frequency f and phase ϕ. Sampling rate is 128 samples/s. We have

$$x_n = \sin\left(\frac{2\pi f n}{128} + \phi\right).$$

At $f = 32$ Hz,

$$x_n = \begin{cases} (-1)^{n/2} \sin\phi & , \quad n \text{ even} \\ (-1)^{(n-1)/2} \cos\phi, & n \text{ odd}. \end{cases}$$

Without loss of generality, we assume that $|\phi| \leq \pi/4$. The output, y_n, of MF_5 is

$$y_n = \begin{cases} -\sqrt{2}\sin\phi \sin\left(\frac{\pi}{2}n + \frac{\pi}{4}\right), & 0 \leq \phi \leq \pi/4 \\ \sqrt{2}\sin\phi \sin\left(\frac{\pi}{2}n - \frac{\pi}{4}\right), & -\pi/4 \leq \phi \leq 0. \end{cases}$$

Thus the amplitude transferred at $f = 32$ Hz is $|\sqrt{2}\sin\phi|$, which ranges between 0 and 1 for $|\phi| \leq \pi/4$. Assuming a uniformly distributed random phase, the average power transferred is

$$\frac{4}{\pi} \int_0^{\pi/4} 2(\sin\phi)^2 \, d\phi = 0.363$$

which is a significant -4.4 dB sidelobe of the power transfer function.

A close look at the output $\{y_n\}$ reveals that it is the recurrent sequence, $\dots, a, a, -a, -a, \dots$ of MF_5 or a fixed point of MF_5^2. In fact, the input $\{x_n\}$ itself is this recurrent sequence provided $\phi = \pm \pi/4$. There is no doubt why there should be such a large sidelobe at this frequency.

Similarly, at $f = 64$ Hz the input is $x_n = \sin(\pi n + \phi) = (-1)^n \sin \phi$, which is a alternating sequence and hence a Type II fixed point of MF_5. Clearly $\{y_n\} = \{x_n\}$ and the transfer function has a peak equal to 1 at this frequency.

To eliminate these undesirable sidelobes, he suggested using MF_4 followed by MF_2, i.e., $MF_4 * MF_2$. Here the output of an even-length median filter is defined by

$$\{y_{n+1/2}\} = MF_{2k}\{x_n\},$$

and

$$y_{n+1/2} = \text{median}(x_{n-k+1}, \dots, x_{n+k}),$$

where the median of an even number of samples is the average of the two middle ones. Therefore, the output y_n of $MF_4 * MF_2$ is

$$y_n = \tfrac{1}{2} \text{median}(x_{n-1}, x_n, x_{n+1}, x_{n+2})$$
$$+ \tfrac{1}{2} \text{median}(x_{n-2}, x_{n-1}, x_n, x_{n+1})$$

which also covers 5 neighboring samples. He showed that the power transfer function of this compound filter has a single sidelobe of -13.3 dB near $f = 43$ Hz. In terms of smoothing, it works far better than the single MF_5 filter; however, step functions are no longer preserved. In fact MF_4 has lost considerably those features of median filtering; as suggested in [6.5], MF_4 should be considered as a 25% trimmed mean filter rather than as a member of the generalized median filters. Also, MF_2 is nothing more than an averaging of each two neighboring samples.

On the other hand, according to our fixed point theory, we may consider the average of MF_3 and MF_5, i.e.,

$$\tfrac{1}{2}MF_3 + \tfrac{1}{2}MF_5.$$

For the sinusoid of $f = 32$ Hz one can easily verify that $MF_3\{x_n\} = -MF_5\{x_n\}$ for arbitrary ϕ; thus $f = 32$ Hz is a null of the transfer function of this filter. Similarly, at $f = 64$ Hz the input, which is an alternating sequence, is a fixed point of MF_5 and a recurrent point of MF_3; by averaging the two, we obtain again a null. Since the filter $0.5 MF_3 + 0.5 MF_5$ does not have any fixed point other than LOMO(4) sequences and we are not aware that it has any recurrent point; thus we believe its transfer function should have much smaller sidelobes than those of MF_3 or MF_5. Indeed, our numerical result bears this out: it shows a maximum sidelobe of -13.0 dB at $f = 51$ Hz and the next highest of -23.3 dB at $f = 37$ Hz.

6.3 Fixed Points of Two-Dimensional Median Filters

To extend the above results on the fixed points of 1-dimensional median filters to their counterparts in a 2-dimensional domain, which is more practical and more interesting in picture processing, one seeks naturally those characteristics that distinguish one type of fixed points from another. The task here is considerably more difficult than the one before. A mere look at the windows involved reveals the problem. In the 1-dimensional case, whenever the window of a median filter is displaced by one step, only a single new sample is introduced and an old one deleted. Therefore, the fixed points of 1-dimensional median filters are greatly restricted in structure. Although we do not know all the properties of Type II fixed points or how to generate all of them and even less do we know about the recurrent points or generalized median filters, we have developed a fairly good idea about their general characteristics. On the contrary, when the window of a 2-dimensional median filter (one that does not degenerate into a line segment) is moved by one step, there is more than one sample introduced or deleted. Intuitively, greater freedom is allowed and fixed points of 2-dimensional median filters will be much more complicated or less structured than their 1-dimensional counterparts. Examples of fixed points give evidence that they can exhibit strong characteristics of Type II fixed points in one region while remain locally monotonic in another; this is not true in the 1-dimensional case. We will return to this issue later.

In Chap. 5 it is shown that edge pictures are preserved under 2-dimensional median filtering if the window is symmetric and includes the center. By its practical importance, only windows of this kind have been considered in the past and will be assumed throughout this chapter. An edge picture is like a step function of the 1-dimensional case; both are the simplest monotonic functions. As noted at the beginning of Sect. 6.1 local monotonicity suffices to make a sequence a fixed point. By analogy, one expects that a picture will be invariant to a median filter of window A so far as it remains *monotonic* within the window as the center of the window is moved from one pixel to another. Clarification of the term *monotonic* is necessary and will be done later. Following an idea of Justusson [6.10] we first decompose a 2-dimensional window into lines and then we make the assumption that whatever that is desired of the window is also true of each line segment. We always assume that the windows are symmetric about and include the origin $(0, 0)$.

Lemma 6.2: Let A be a window and let L be an arbitrary line in R^2. If

$$\text{median}(x_{i,j}|(i,j)\in L\cap A)=x_{0,0} \tag{6.5}$$

for all L passing through the origin, then $\text{median}(x_{i,j}|(i,j)\in A)=x_{0,0}$.

Proof: From the assumption that A is symmetric about and includes $(0, 0)$, $L\cap A$ must have an odd number of points in it and, excluding $x_{0,0}$, half of the samples are $\geq x_{0,0}$ and the other half $\leq x_{0,0}$ in order that (6.5) holds. Since all the lines which pass through the origin are nonoverlapping except at the origin, excluding $(0,0)$ half of the samples in A must be $\geq x_{0,0}$ and the other half $\leq x_{0,0}$.

For a window A we shall say that a picture $\{x_{i,j}\}$ is *locally monotonic with respect to A* if, for all (r, s), the relative translation of the picture with respect to the window, and for all lines L which pass through $(0,0)$, the center of the window, $\{x_{i+r,j+s}\}$ is monotonic on the restriction of L to the window A.

Lemma 6.3: If a picture $\{x_{i,j}\}$ is locally monotonic with respect to a window A, then it is a fixed point of the median filter with a window equal to A or being a subset of A.

Proof: By assumption, for any (r, s), $\{x_{i+r,j+s}\}$ is monotonic on $L \cap A$ for all L passing through the origin; hence

$$\text{median}(x_{i+r,j+s}|(i, j) \in A \cap L) = x_{r,s}.$$

A simple application of Lemma 6.2 completes the proof.

To further relax the requirement on local monotonicity in the above lemma, we may consider the following class of windows.

Definition: A window A is *p-symmetric* if, in addition to being symmetric about and including $(0,0)$, it contains all the points (r, s) of the intersection of $\mathbb{Z}^2$ and the finite line segment $\theta(i, j)$, $0 < \theta < 1$, for all (i, j) in A. Note that $\theta(i, j)$ is the line segment linking $(0,0)$ and (i, j).

Now let L be an arbitrary line in $\mathbb{Z}^2$ and let A be a *p*-symmetric window. Then the points contained in L are located periodically. For a point (r, s) on L, we denote by $N_{L,A}$ the number of points contained in $L \cap \{A + (r, s)\}$, where $\{A + (r, s)\}$ denotes the window A shifted so that its center is at (r, s). By the periodicity of $\mathbb{Z}^2$ the number $N_{L,A}$ is independent of the particular (r, s) chosen, and by the symmetry of A it is odd. In fact, all lines in $\mathbb{Z}^2$ and parallel to L have the same $N_{L,A}$. The following theorem, which extends Theorem 6.1 to the 2-dimensional domain, is a generalization of a result of *Justusson* [6.10] by using the idea of locally monotonic sequences.

Theorem 6.8: Let A be a *p*-symmetric window and let $\{x_{i,j}\}$ represent a picture. If for every line L, the samples $x_{i,j}$ on it are locally monotonic of length $(N_{L,A} + 3)/2$, then $\{x_{i,j}\}$ is invariant to the median filter of a *p*-symmetric window which is equal to A or a subset of A.

Proof: Since the samples on a line L with L passing through an arbitrary point (r, s) are locally monotonic of length $(N_{L,A} + 3)/2$, by Theorem 6.1 they are invariant to 1-dimensional median filtering of window length $\leq N_{L,A}$. For any p-symmetric window B which is equal to or a subset of A, $N_{L,B} \leq N_{L,A}$ and the intersection $L \cap \{B+(r,s)\}$ is a subsegment, centered at (r,s), of consecutive points in L. Therefore,

$$\text{median}(x_{i+r,j+s}|(i,j) \in B, (i+r,j+s) \in L) = x_{r,s}.$$

By Lemma 6.2 the theorem holds.

Similar to Theorem 6.4, a converse to the above theorem exists. The proof is the same as that of Theorem 6.4 and therefore is omitted.

Theorem 6.9: If a picture $\{x_{i,j}\}$ is invariant to all median filters of p-symmetric window which is equal to A or a subset of A, where A is p-symmetric, then for any line L the samples $x_{i,j}$ on it are locally monotonic of length $(N_{L,A}+3)/2$.

If a picture is invariant to all median filters of p-symmetric window, then by the above theorem its restriction on any line L in $\mathbb{Z}^2$ has to be monotonic. Pictures possessing this property are called *monotonic in all directions* [6.10] and it turns out that they have a simple structure; the following result is due to *Justusson* [6.10].

If $\{x_{i,j}\}$ is monotonic in all directions, then there exists a slope b such that $x_{i,j} \geq x_{r,s}$ for all $j-bi > s-br$, or $x_{i,j} \leq x_{r,s}$ for all $j-bi > s-br$. In case $b = \pm \infty$, $j-bi > s-br$ is replaced with $i > r$. The slope b is unique unless $\{x_{i,j}\}$ is a constant.

An outline of the proof is given below. Consider the level function $g_c(x)$ of (6.2) and the picture $\{g_c(x_{i,j})\}$. Let $S_{1,c} = \{(i,j)|g_c(x_{i,j})=1\}$ and $S_{0,c}$ the complement of $S_{1,c}$ in $\mathbb{Z}^2$. Obviously, $S_{1,c} \supset S_{1,d}$ for $c < d$. Suppose that $S_{1,c}$ and $S_{0,c}$ are both nonempty. Then by the assumption $\{x_{i,j}\}$ being monotonic in all directions one can show that both are convex, i.e., for any point in $\mathbb{Z}^2$ which is a convex combination of points of $S_{1,c}$ is also in $S_{1,c}$. Furthermore, they are *exclusive* in the sense that the minimal convex sets in $\mathbb{R}^2$, denoted by $C(S_{1,c})$ and $C(S_{0,c})$, of $S_{1,c}$ and $S_{0,c}$ respectively are exclusive. The union $C(S_{1,c}) \cup C(S_{0,c})$ may not be equal to $\mathbb{R}^2$. Then by the principle of separating hyperplanes, there is a line $y = a+bx$ which separates $C(S_{1,c})$ and $C(S_{0,c})$. [$C(S_{1,c})$ and $C(S_{0,c})$ both may have points on $y = a+bx$.] The slope is unique even though the separating line may not be. Clearly, $x_{i,j} \geq c$ for all $j-bi > a$ and $x_{i,j} < c$ for all $j-bi < a$ or vice versa. Since $S_{1,c} \supset S_{1,d}$ provided $c < d$, one can show that b is unique for all c unless $S_{1,c}$ or $S_{0,c}$ is empty. The proof is completed by considering all $g_c(x)$. By the above result, if a line L is not parallel to $y = bx$ and if $x_{i_k,j_k} = c$ for all $k = 1, 2, 3$, where (i_k, j_k) are three consecutive points on L, then $x_{i,j} = c$ for all (i,j) on the line L^* which is parallel to $y = bx$ and intersects L at (i_2, j_2).

The sufficient conditions given in Lemma 6.3 or in Theorem 6.8 usually are difficult to meet, especially, for large windows. In Fig. 6.1 some binary pictures which satisfy the sufficient condition of Theorem 6.8 (which is less stringent than that of Lemma 6.3) are given for several p-symmetric windows. These

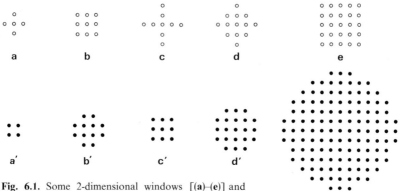

Fig. 6.1. Some 2-dimensional windows [(a)–(e)] and examples of their respective fixed points [(a')–(e')]

binary pictures can be considered either as uniform objects against a uniform white background or as outputs of level functions $g_a(\cdot)$. They are constructed as follows.

For each window, consider those directed line segments (spokes) which emanate from the center (base) and end at a point (tip) on the boundary. Each of these line segments can also be identified by its angle θ, $0° \leq \theta \leq 360°$ with respect to a segment joining the center, say, $(0, 0)$ and the point $(1, 0)$. Then one can easily recognize that each object has a boundary which is piecewise-linear and is made up of those spokes in the descending or ascending order of θ, each with its base connected to the tip of its predecessor. Then the object is the convex set in $\mathbb{Z}^2$ with the above constructed boundary. Not all the binary pictures constructed in this way are locally monotonic in the sense of Theorem 6.8. For example, if the "cross" sign, which consists of the five points $(0, 0)$ and (i, j) where $|i| = |j| = 1$, is used as the window, then the object so constructed has boundary points $(1, 0)$, $(0, 1)$, $(-1, 0)$, $(0, -1)$ and local monotonicity in the sense of Theorem 6.8 does not hold at the center of the object. However, it seems that it is true provided that the window A satisfies the following;

$$\lim_{n \to \infty} A_n = \mathbb{Z}^2 ,$$

where $A_{n+1} = A_n + A = \{x + y | x \in A_n$ and $y \in A\}$ and $A_1 = A$. We can consider a window A as *degenerate* if it fails to satisfy the above condition, because, in this case the 2-dimensional domain $\mathbb{Z}^2$ can be decomposed into disjoint regions which are cosets of A^* where $A^* = \lim_{n \to \infty} A_n$, and median filtering with the degenerate window A can be equally accomplished by applying it separately to each of the cosets of A^*. There is a total lack of interaction among cosets of A^*.

Returning to Fig. 6.1 we can make the following observations: Firstly, one can see that none of the windows is degenerate. Secondly, each object is the smallest *finite* convex object which is preserved under median filtering of the repective window. Even though we have put the restriction that each line L be

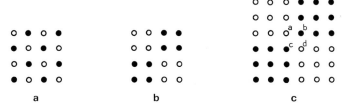

Fig. 6.2a–c. Examples of fixed points of the 3 ×3 square window

locally monotonic and be invariant to $MF_{N_{L,A}}$, it appears that in the case of *finite* convex object the sufficient condition of Theorem 6.8 is also necessary in order that it is preserved under median filtering. Finally, by the way we have constructed the boundary of each convex object, we are led to the conjecture that, in general, for a binary picture of smooth objects against a uniform background to be invariant to median filtering with a non-degenerate *p*-symmetric window, the following might be necessary or sufficient: that the *convex* part of the boundaries of the objects or the background is composed of spokes of equal or greater lengths of the window and that the corresponding spokes of any two connected linear segments of the contour are also adjacent spokes of the window. This would put some restrictions on the contours of preservable pictures. Unfortunately, the condition is neither necessary nor sufficient. In fact, terms like boundary or linear segments of boundary are rather loosely defined for general binary pictures.

At the beginning of this section it was mentioned that unlike the fixed points of MF_{2k+1}, which are either $LOMO(k+2)$ or nowhere $LOMO(k+1)$, fixed points of 2-dimensional median filters can be of a mixed nature. Some examples are shown in Fig. 6.2. Consider the periodic extensions of the patterns (a)–(c) of Fig. 6.2. Obviously, they are the same except with different scales and all three are invariant to median filtering of the 3 × 3 square window, which is denoted by A. If *smoothness* is judged by local monotonicity with respect to the 3 × 3 square window (see definition following Lemma 6.2), then both (a) and (b) are *nowhere* locally monotonic with respect to A. Pattern (c) is locally monotonic with respect to A except at those *saddle points* such as **a**, **b**, **c**, and **d** where it fails. With an even smaller scale the pattern, which remains a fixed point of the median filter of window A, becomes smoother; however, those saddle points have also persisted. Therefore one may consider pattern (c) as a mixture-type fixed point.

6.4 A Fast Median Filtering Algorithm

An interesting and efficient algorithm for doing 2-dimensional median filtering of arbitrary window was given in *Huang* et al. [6.4]. It makes use of the fact that only a portion of the samples contained in the window is deleted and an equal number of new samples added as the window is displaced by one unit. This section is based on [6.4].

Even though this algorithm works for arbitrary windows, we shall use as an example a rectangular window of size $m \times n$, where m is the number of columns. The window is assumed to sweep from left to right horizontally and back to the next row. Boundary points are not considered; they may be treated with windows reduced in size as in the 1-dimensional case. As usual, the picture is quantized to, say, 256 gray levels. Then the following three quantities are computed for those picture elements (pels) of the first (initial) window and updated at each step to the right:

1) hist, the gray level histogram;
2) mdn, the median;
3) ltmdn, number of pels having gray level *less* than mdn.

As the window moves to the right by one unit, each pel, g, of the leftmost column of the previous window is deleted. The histogram and the count, ltmdn, are updated by

$$\text{hist}[g] \leftarrow \text{hist}[g] - 1$$

and

$$\text{ltmdn} \leftarrow \text{ltmdn} - 1 \quad \text{if} \quad g < \text{mdn}. \tag{6.6}$$

Similarly, each pel, g, of the rightmost column of the current window is added. We update hist and ltmdn accordingly by

$$\text{hist}[g] \leftarrow \text{hist}[g] + 1$$

and

$$\text{ltmdn} \leftarrow \text{ltmdn} + 1 \quad \text{if} \quad g < \text{mdn}. \tag{6.7}$$

Upon completion, hist gives the histogram of the current window and ltmdn is the count of pels of the current window having gray level less than the median of the previous window.

Next we find the median of the current window by moving down (or up) the mdn according to whether ltmdn is greater than (or not greater than) $(mn - 1)/2$, which is denoted by th. First, we

$$\text{compare ltmdn with th}. \tag{6.8}$$

Then consider the following two cases:

Case I. ltmdn > th, which indicates that mdn is greater than the median of the current window. Update by

$$\text{mdn} \leftarrow \text{mdn} - 1$$

$$\text{ltmdn} \leftarrow \text{ltmdn} - \text{hist}[\text{mdn}]$$

until

$$\text{ltmdn} \leq \text{th}. \tag{6.9}$$

Case II. ltmdn $\leq$ th, which indicates that mdn is less than or equal to the median of the current window. Test

$$\text{ltmdn} + \text{hist}[\text{mdn}] \leq \text{th}. \tag{6.10}$$

If no, then mdn is exactly the median of the current window. If yes, which indicates that mdn + 1 is still less than or equal to the desired median, then update

ltmdn $\leftarrow$ ltmdn + hist[mdn]

mdn $\leftarrow$ mdn + 1

and go back to (6.10).

Clearly, (6.6) and (6.7) each take up n comparisons. Let d = median of current window – median of previous window. For Case I, i.e., $d < 0$, (6.8) and (6.9) together require $1 + |d|$ comparisons; for Case II, $d \geq 0$, (6.8) and (6.10) together require $2 + d$ comparisons. Let $P_i = \text{Prob}(d = i)$. Then the expected number of comparisons is

$$\bar{c} = 2n + \sum_{i<0}(1 + |i|)P_i + \sum_{i \geq 0}(2 + i)P_i$$
$$= 2n + \overline{|d|} + 1 + \text{Prob}(d \geq 0),$$

where $\overline{|d|}$ is the expected value of $|d|$. Assuming $\text{Prob}(d > 0) = \text{Prob}(d < 0)$, we have

$$\bar{c} = 2n + \overline{|d|} + 1.5 + 0.5 P_0.$$

Huang et al. [6.4] indicated that $\overline{|d|}$ is usually small, and experimental results in the same report show that it is less than 10. Because of the small $\overline{|d|}$, considerable saving in computation time is possible even for small $n \times m$. As the window grows in size, this algorithm becomes much superior to those without taking advantage of the fact that the current window has $n(m-2)$ pels common with those of its predecessor.

6.5 Conclusions

In this chapter we have presented some recent results on the properties of median filters, especially on their fixed points and the relationship between fixed points and experimental observations previously reported by other researchers. At the end, an interesting algorithm due to *Huang* et al. is also included.

By definition, fixed points are sequences which are invariant to their respective median filters and hence the influence of median filtering on them is none and clearly understood. On the other hand, for those sequences which are neither fixed points nor recurrent points the effect of median filtering, without a thorough statistical investigation, can only be qualitatively discussed. It would be misleading if the merit of a median filter or a generalized median filter is judged exclusively by its fixed points or recurrent points. Here Chap. 5 fills the need.

Type II-like fixed points or recurrent points of 2-dimensional median filters are not studied; only a few examples are given in Fig. 6.2. It is not clear whether they pose any problem in real-world picture processing. If they do, then one may be able to overcome it by considering weighted median filters which have been shown in Sect. 6.2 to be effective for the 1-dimensional case.

Appendix 6.A

In this appendix, we prove Theorems 6.2 and 6.3. We need the following lemma.

Lemma 6.A.1: Let $n<m$ and let $x_n<x_i<x_m$ for all $n<i<m$. If median $(x_{n-p}, ..., x_{n+p})=x_n$ and if median $(x_{m-q}, ..., x_{m+q})=x_m$, where $n-p\leq m-q$ and $n+p\leq m+q$, then $x_j\geq x_m$ or $x_j\leq x_n$ for each j, where $n-p\leq j<n$ or $m<j\leq m+q$. Furthermore, $x_j\leq x_n$ for all $n-p\leq j<\min(m-q, n)$ and $x_j\geq x_m$ for all $\max(m, n+p)<j\leq m+q$.

Proof: If the median of $(x_{n-p}, ..., x_{n+p})=x_n$, then out of this segment there are at least p samples excluding x_n which are $\leq x_n$. Similarly, out of $(x_{m-q}, ..., x_{m+q})$ there are at least q samples other than x_m which are $\geq x_m$. Since $x_n<x_i<x_m$ for all $n<i<m$ and since it is assumed that $n-p\leq m-q$ and $n+p\leq m+q$, we have at least p samples $\leq x_n$ and at least q samples $\geq x_m$ out of the segments $(x_{n-p}, ..., x_{n-1})$ and $(x_{m+1}, ..., x_{m+q})$, which contain only $p+q$ samples when combined. Therefore, of the $p+q$ samples there are exactly p samples $\leq x_n$ and q samples $\geq x_m$. This proves the first part. To prove the second part, we have only to note that if there exists a j, where $n-p\leq j<\min(m-q, n)$, such that $x_j\geq x_m$, then excluding x_m we have at most $q-1$ samples which are $\geq x_m$ out of $(x_{m-q}, ..., x_{m+q})$ because of the first part of the lemma. This contradicts the assumption that $x_m=$ median $(x_{m-q}, ..., x_{m+q})$.

Proof of Theorem 6.2: Assume that the segment $(x_{n-k}, ..., x_n)$, which has $k+1$ samples, is monotonic and increasing, i.e., $x_{n-k}<x_n$; otherwise $(x_{n-k}, ..., x_{n+1})$ is monotonic whatever x_{n+1} is, if $x_{n-k}=x_n$. If $x_{n+1}\geq x_n$, then $(x_{n-k}, ..., x_{n+1})$, which has $k+2$ samples, is monotonic and we can proceed to $(x_{n-k+1}, ..., x_{n+1})$, which is monotonic, and examine x_{n+2}. If $x_{n+1}<x_n$, then by Lemma 6.A.1 we must have $x_{n-k}\geq x_n$ which is contrary to the assumption that $x_{n-k}<x_n$; thus x_{n+1} must be $\geq x_n$. The same argument can be applied to x_{n-k-1} to show that $x_{n-k-1}\leq x_{n-k}$ and hence $(x_{n-k-1}, ..., x_n)$ is also monotonic. The proof is completed by applying the argument to $x_{n+2}, x_{n+3}, \cdots$ and $x_{n-k-2}, x_{n-k-3}, \cdots$.

Proof of Theorem 6.3: Assume that $\{x_n\}$ is a fixed point of MF_{2k+1} and that it is *nowhere* $\text{LOMO}(k+1)$. We shall investigate all the possibilities around a transition, say, at the origin. Without loss of generality we assume that:

I) $x_0 > x_1$.

Then by Lemma 6.A.1 we have

II) $x_i \geq x_0$ or $x_i \leq x_1$ for each i, where $-k \leq i < 0$ or $1 < i \leq k+1$.

In addition, we have

III) $x_{-k} \geq x_0$ and $x_{1+k} \leq x_1$.

Consider the following case:

IV) $x_i \geq x_0$ for all $-k \leq i < 0$. Let j be the smallest integer satisfying

$$x_j > x_{j+1} \geq \ldots \geq x_0 > x_1, \qquad -k \leq j \leq -1. \tag{6.A.1}$$

If j does not exist, then we must have $x_{-k} = x_{-k+1} = \ldots = x_0$. If j exists, then by Lemma 6.A.1 and the assumption $x_i \geq x_0$ we have $x_i \geq x_j$ for all $-k \leq i \leq j$; in fact, $x_i = x_j$ because j is the smallest integer satisfying (6.A.1). Therefore $(x_{-k}, x_{-k+1}, \ldots, x_0)$ is monotonic of length $k+1$ which violates the assumption than x_n is *nowhere* $\text{LOMO}(k+1)$. We conclude that
  a) there is at least one sample, which is $\leq x_1$ out of $(x_{-k+1}, \ldots, x_{-1})$, and similarly,
  b) there is at least one sample which is $\geq x_0$ out of $(x_2, \ldots, x_k)$.
It suffices to consider only IVa):
  V) There exists i, $-k+1 \leq i \leq -1$, such that $x_i \leq x_1$.
If there exists i, $x_i < x_1$, then let p be the greatest integer such that $-k+1 \leq p \leq -1$ and $x_p < x_1$. Furthermore, let j be the least integer such that $p < j \leq 0$ and $x_j \geq x_0$. Clearly, $x_p < x_{p+1} = \ldots = x_{j-1} < x_j$ by II). By Lemma 6.A.1 we have either $x_n \geq x_j$ or $x_n \leq x_p$ for every n, $j < n \leq j+k$. Since $j < 1 < j+k$, either $x_1 \geq x_j > x_0$ or $x_1 \leq x_p < x_1$. This contradicts our assumptions. We conclude that

  a) $x_i = x_1$ if $x_i \leq x_1$, where $-k+1 \leq i \leq -1$, and by analogy,
  b) $x_n = x_0$ if $x_n \geq x_0$, where $2 \leq n \leq k$.

Next we consider
  VI) Let q be the greatest integer such that $-k \leq q \leq -1$, $x_q \neq x_1$, and $x_q \neq x_0$. By II) and Va), we have $x_q > x_0$. Let m be the least integer such that $q < m \leq 1$ and $x_m = x_1$. Then $x_q > x_{q+1} = \ldots = x_{m-1} > x_m$. By Lemma 6.A.1, we have $x_n \geq x_q$ or $x_n \leq x_m$ for every n, $m < n \leq m+k$. Consider the following two possibilities:
  a) If $m < 0$, then the above implies that $x_0 \geq x_q > x_0$ or $x_0 \leq x_m = x_1$. Neither is true.

b) If $m=1$, then we have $x_n \geq x_q > x_0$ or $x_n \leq x_m = x_1$ for every n, $1 < n \leq 1+k$. By IVb) and Vb), there must exist at least one sample x_i, $1 < i \leq k$, such that $x_i = x_0$ which is contradictory to the above conclusion.

We conclude that such q does not exist, or equivalently, $x_i = x_1$ or $x_i = x_0$ for every i, $-k \leq i \leq -1$. By analogy, the above also holds for all samples x_i, $2 \leq i \leq k+1$. By IVa) there is at least one transition in the segment $(x_{-k}, \ldots, x_0)$, and similarly by IVb) a transition in the segment $(x_1, \ldots, x_{1+k})$. The proof is completed by applying the same argument to those transitions in the two above-mentioned segments and beyond.

Appendix 6.B

Theorem 6.5 is proved in this appendix and it is also shown that the theorem is a special case of some properties belonging to a class of more general smoothers which are similar to MF_3. We shall first discuss this class in general.

If we let $\{z_n\} = T\{x_n\}$ where T is the operator under consideration and let $y_n = \text{median}(x_{n-1}, x_n, x_{n+1})$, then there are three constraints of increasing strictness which we may impose on T:

i) $(z_n - x_n)(z_n - y_n) \leq 0$ for all n.
ii) In addition to i), $y_n \leq z_n < x_n$ or $y_n \geq z_n > x_n$, if $x_n \neq y_n$.
iii) In addition to i), $(z_n - x_n)(z_n - y_n) < 0$, if $x_n \neq y_n$.

In other words, if $x_n \neq y_n$, then z_n is contained in i) the closed interval of x_n and y_n, ii) the half open interval excluding x_n, or iii) the open interval, respectively. For example, MF_3^2 satisfies i) only, MF_3 satisfies ii) but not iii), and $T = 1/2 I + 1/2 MF_3$ satisfies iii). Here I stands for the identity operator. We have the following simple lemma:

Lemma 6.B.1: Suppose that T satisfies i). If (x_{n-1}, x_n, x_{n+1}) is monotonic, then $z_n = x_n$ and (z_{n-1}, z_n, z_{n+1}) is monotonic of the same trend.

The proof is straightforward and is omitted. Obviously, LOMO(3) sequences are fixed points of any T satisfying i). The next lemma shows that some properties are preserved under linear combination or compounding of two smoothers.

Lemma 6.B.2: Let $0 < \alpha < 1$ and let $T_1 * T_2$ be the compound smoother $T_2\{T_1\{x_n\}\}$. The following are true:
a) If T_1 and T_2 satisfy i), then both $\alpha T_1 + (1-\alpha)T_2$ and $T_1 * T_2$ satisfy i).
b) If T_1 satisfies ii) and T_2 satisfies i), then $\alpha T_1 + (1-\alpha)T_2$ satisfies ii).
c) If T_1 satisfies iii) and T_2 satisfies i), then $\alpha T_1 + (1-\alpha)T_2$ satisfies iii) and $T_1 * T_2$ satisfies ii). Furthermore, if T_2 also satisfies iii), then $T_1 * T_2$ satisfies iii).

Remark: Even if both T_1 and T_2 satisfy ii), the compound $T_1 * T_2$ may not satisfy ii).

Proof: The proof related to the convex combinations is trivial; therefore we shall consider only the compound cases. Let

$$T_1\{x_n\} = \{t_n\}, T_2\{t_n\} = \{s_n\} \quad \text{and} \quad MF_3\{x_n\} = \{y_n\}.$$

If $y_n = x_n$, then $t_n = x_n$ and (t_{n-1}, t_n, t_{n+1}) is monotonic by Lemma 6.B.1; hence $s_n = t_n$. If $y_n \neq x_n$, say $x_n < y_n = x_{n-1}$, then we consider a) and c) separately:

a) We have $x_n \leq y_{n-1} \leq t_{n-1} \leq x_{n-1} = y_n$. Since median$(t_{n-1}, t_n, t_{n+1})$ is between t_{n-1} and t_n, where both are between x_n and y_n, s_n is contained in the closed interval of x_n and y_n.

c) We have either $y_{n-1} = x_{n-1}$ or $x_n \leq y_{n-1} < x_{n-1}$. Therefore, either $t_{n-1} = x_{n-1}$ or $x_n < t_{n-1} < x_{n-1}$ and the median(t_{n-1}, t_n, t_{n+1}), which is between t_{n-1} and t_n, is in the half open interval $(x_n, y_n]$. Thus $x_n < s_n \leq y_n$. If T_2 satisfies iii) also, then $x_n < s_n < y_n$.

We have pointed out before that MF_3 satisfies ii); in fact, any smoother satisfying ii) has LOMO(3) sequences as its only fixed points.

Lemma 6.B.3: If T satisfies ii) and if $\{x_n\}$ is a fixed point of T, then it is a LOMO(3) sequence.

Proof: If $\{x_n\}$ is not LOMO(3), then there exists n such that $y_n \neq x_n$, say, $y_n > x_n$. By assumption T satisfies ii); thus $z_n > x_n$, where $\{z_n\} = T\{x_n\}$. This is contradictory to the assumption that $\{x_n\}$ is a fixed point of T.

For compound smoothers, we have the following lemma:

Lemma 6.B.4: Let $T = T_1 * T_2$ and let $\{x_n\}$ be a fixed point of T. Then
a) $\{x_n\}$ is either a LOMO(3) sequence or an alternating sequence (i.e., ..., 0, 1, 0, 1, ...) if T_1 satisfies ii) and T_2 satisfies i),
b) $\{x_n\}$ is a LOMO(3) sequence if T_1 satisfies iii) instead. The above results hold for $T = T_2 * T_1$.

Proof: Use the notations defined in the proof of Lemma 6.B.2. Suppose that $\{x_n\}$ is a fixed point of T; however, it is not LOMO(3). Then there exists n such that $y_n \neq x_n$. Assume that $x_n < y_n$.

a) If T_1 satisfies ii), then $x_n < t_n \leq y_n$, $x_n \leq t_{n-1} \leq x_{n-1}$, and $x_n \leq t_{n+1} \leq x_{n+1}$. In order that $s_n = x_n$ we must have $x_n = $ median(t_{n-1}, t_n, t_{n+1}) or $x_n = t_{n-1} = t_{n+1}$. This implies that $x_n \leq s_{n+1} \leq t_n$ and $x_n \leq s_{n-1} \leq t_n$. Hence $t_n = y_n = x_{n-1} = x_{n+1}$. Since $t_{n-1} = x_n < x_{n-1}$, (x_{n-2}, x_{n-1}, x_n) cannot be monotonic and the same is true for (x_n, x_{n+1}, x_{n+2}). By the above argument $x_{n-2} = x_n = x_{n+2}$; thus $\{x_n\}$ must be an alternating sequence. If T_1 satisfies i) and T_2 satisfies ii) instead, then (t_{n-1}, t_n, t_{n+1}) cannot be monotonic, otherwise (s_{n-1}, s_n, s_{n+1}) is also monotonic thus violating the assumption that $y_n \neq x_n$. If median$(t_{n-1}, t_n, t_{n+1}) > t_n$, then $s_n > t_n \geq x_n$, which cannot be true because $\{x_n\}$ is a fixed point of T. Therefore we have median$(t_{n-1}, t_n, t_{n+1}) < t_n$ and hence $s_n = t_{n-1} = t_{n+1} = x_n$; this implies that $\{x_n\}$ is an alternating sequence.

b) If T_1 satisfies iii) and T_2 satisfies i), then by Lemma 6.B.2c), $T = T_1 * T_2$ satisfies ii) and by Lemma 6.B.3 it has only LOMO(3) sequences as its fixed points. If, instead, T_1 satisfies i) and T_2 satisfies iii), then (t_{n-1}, t_n, t_{n+1}) cannot be monotonic, otherwise (s_{n-1}, s_n, s_{n+1}) is monotonic thus violating the assumption that $y_n \neq x_n$. If median$(t_{n-1}, t_n, t_{n+1}) > t_n$, then $s_n > t_n \geq x_n$ which is impossible because $\{x_n\}$ is a fixed point. Finally, if median$(t_{n-1}, t_n, t_{n+1}) < t_n$, then we have $x_n \leq$ median$(t_{n-1}, t_n, t_{n+1}) < s_n < t_n$ which contradicts the fixed point assumption again. Therefore, we conclude that $\{x_n\}$ must be LOMO(3).

Remark: As mentioned above, MF_3 has LOMO(3) sequences as its only fixed points. However, a direct application of Lemma 6.B.4a) shows that MF_3^k, which can be written as $MF_3 * MF_3^{k-1}$, may have both LOMO(3) and alternating sequences as its fixed points. One can easily check that the alternating sequences are fixed points of MF_3^k if and only if k is even. For odd k, it has only LOMO(3) fixed points. Consequently, the alternating sequences are the only *recurrent* points of MF_3.

Since an alternating sequence is usually considered rather rough, it is unwise to use MF_3 or MF_3^k as the only smoother in smoothing. In the following we consider a simple arrangement which can forestall this unwanted sequence. First we need a lemma.

Lemma 6.B.5: Let $0 < \alpha < 1$ and let T_1 and T_2 satisfy i). Then $\{x_n\}$ is a fixed point of $T = \alpha T_1 + (1-\alpha)T_2$ if it is a fixed point of T_1 and T_2 simultaneously.

Proof: Trivial.

Combining the above lemma and the remark following Lemma 6.B.4 we have

Lemma 6.B.6: Let $a_0 \neq 1$, $a_k \geq 0$, and $\sum_{k=0}^{n} a_k = 1$. Then the smoother

$$T = \sum_{k=0}^{n} a_k MF_3^k$$

has fixed points LOMO(3) sequences only provided for some odd k, $a_k \neq 0$. If $a_k = 0$ for all odd k, then T has both LOMO(3) and alternating sequences as its only fixed points.

To obtain the fixed points of T^m we have only to rewrite T^m as follows:

$$T^m = T^{m-1} * \left(\sum_{k=0}^{n} a_k MF_3^k \right)$$

$$= a_0 T^{m-1} + a_1 T^{m-1} * MF_3 + \sum_{k=2}^{n} a_k T^{m-1} * MF_3^{k-1} * MF_3.$$

If $\{x_n\}$ is a fixed point of T^m, then by Lemmas 6.B.4 and 6.B.5 and the assumption that $a_0 \neq 1$, it must be a LOMO(3) or an alternating sequence. We can check that the alternating sequence cannot be invariant to T^m unless 1) $a_k = 0$ for all even k and m is even, or 2) $a_k = 0$ for all odd k. In the second case, the alternating sequence is a fixed point of T.

References

6.1   J.W.Tukey: *Exploratory Data Analysis* (Addison-Wesley, Reading, MA 1971)
6.2   L.R.Rabiner, M.R.Sambur, C.E.Schmidt: IEEE Trans. ASSP-**23**, 552–557 (1975)
6.3   B.R.Frieden: J. Opt. Soc. Am. **66**, 280–283 (1976)
6.4   T.S.Huang, G.J.Yang, G.Y.Tang: "A Fast Two-Dimensional Median Filtering Algorithm"; School of Electrical Engineering, Purdue University (1977)
6.5   P.F.Velleman: "Robust Non-Linear Data Smoothing"; Tech. Rpt. 89, Ser. 2, Dept. of Statistics, Princeton University (1975)
6.6   J.W.Tukey: *Exploratory Data Analysis* (Addison-Wesley, Reading, MA 1977)
6.7   P.F.Velleman: Proc. Nat. Acad. Sci. USA **74**, 434–436 (1977)
6.8   J.W.Tukey: Class Notes on Special Topics in Statistics, Statistics 411, Dept. of Statistics, Princeton University (1974)
6.9   I.J.Schoenberg: "Some Analytical Aspects of the Problem of Smoothing", in *Studies and Essays Presented to R. Courant on his 60th Birthday, January 8, 1948* (Interscience, New York 1948) pp. 351–370
6.10 B.Justusson: "Median Filters on Deterministic Signals"; report in preparation, Math. Inst., Royal Institute of Technology, Stockholm, Sweden (1979)

Additional References with Titles

Transforms

B. Arambepola: General discrete Fourier transform and the fast Fourier algorithm. Proc. EUSIPCO-80, Sept. 16–18, 1980, Lausanne, Switzerland, pp. 583–588

B. Arambepola, P. J. W. Rayner: Discrete transforms over residue class polynomial rings with applications in computing multidimensional convolutions. IEEE Trans. ASSP-**28** (4), 407–414 (1980)

J. W. Cooley, S. Winograd: A limited range DFT algorithm, Proc. ICASSP-80, April 9–11, 1980, Denver, Colorado, pp. 213–217

F. A. Kamagar, K. R. Rao: Fast algorithms for the 2-D discrete cosine transformation. Proc. IEEE Intern. Symp. on Circuits and Systems, April 28–30, 1980, pp. 206–209

J. Makhoul: A fast cosine transform in one and two dimensions. IEEE Trans. ASSP–**28** (1), 27–34 (1980)

D. C. Munson, B. Lin: Floating point error bound in the prime factor FFT. Proc. ICASSP-80, April 9–11, 1980, Denver, Colorado, pp. 69–72

H. Nawab: "Parallelism and code optimization issues in FFT and WFTA algorithms"; S. M. thesis, Massachusetts Inst. Technol., Cambridge (1978)

H. Nawab, J. H. McClellan: Corrections to "Bounds on the minimum number of data transfers in WFTA and FFT programs". IEEE Trans., ASSP-**28**, 480–481 (1980)

H. J. Nussbaumer: Fast polynomial transform algorithms for digital convolution. IEEE Trans. ASSP-**28** (2), 205–215 (1980)

H. J. Nussbaumer: Fast polynomial transform methods for multidimensional DFTs. Proc. JCASSP-80, April 9–11, 1980, Denver, Colorado, pp. 235–237

R. W. Patterson: "Fixed point error analysis of the Winograd Fourier transform algorithm"; M. S. thesis, Massachusetts Inst. Technol. Cambridge (1977)

B. Rice: Some good fields and rings for computing number theoretic transforms. IEEE Trans. ASSP-**27** (4), 432–433 (1979)

H. F. Silverman: A method for programming the complex, general-N Winograd Fourier transform algorithm. Proc. ICASSP-77, Hartford, CT, May 1977, pp. 369–372

S. Zohar: "Outline of a fast hardware implementation of Winograd's DFT algorithm", in IEEE 1980 Intern. Conf. ASSP-**3**, 796–799

Median Filters

E. Ataman, V. K. Aatre, K. M. Wong: Some statistical properties of median filters, Tech. Rpt. Dept. of Electrical Engineering, Nova Scotia Technical College, Halifax, Canada (1979)

E. Ataman, V. K. Aatre, K. M. Wong: A fast method for real-time median filtering. IEEE Trans. ASSP-**28** (4), 415–421 (1980)

G. Heygster: „Rangordnungsoperatoren in der digitalen Bildverarbeitung", Ph. D. Thesis, Math. Nat. Fakultät, Georg-August-Universität, Göttingen, (1979)

G. Heygester: Rank filters in digital image processing, Proc. 5th Intern. Conf. on Pattern Recognition, Dec. 1–4, 1980, Miami Beach, Florida, pp. 1165–1167

T. Kitahashi, O. Saito, L. Abele, F. Wahl, H. Marko: An extension of median filtering and its deterministic properties, Proc. 5th Intern. Conf. on Pattern Recognition, Dec. 1–4, 1980, Miami Beach, Florida, pp. 1177–1179

C. L. Mallows: "Some theoretical results on Tukey's 3 R smoothers", in *Smooting Techniques for Curve Estimation*, ed. by Th. Gasser, M. Rosenblatt, Springer Lecture Notes in Mathematics 757, (Springer, Berlin, Heidelberg, New York, 1979) pp. 77–90

Subject Index

T. Kohonen

Content-Addressable Memories

1980. 123 figures, 36 tables. XI, 368 pages
(Springer Series in Information Sciences, Volume 1)
ISBN 3-540-09823-2

Contents:
Associative Memory, Content Addressing, and Associative Recall. –
Content Addressing by Software. – Logic Principles of Content-
Addressable Memories. – CAM Hardware. – The CAM as a System
Part. – Content-Addressable Processors. – References. – Subject
Index.

H. J. Nussbaumer

Fast Fourier Transform and Convolution Algorithms

1980. 34 figures, 38 tables. 264 pages
(Springer Series in Information Sciences, Volume 2)
ISBN 3-540-10159-4

Contents:
Introduction. – Elements of Number Theory and Polynomial
Algebra. – Fast Convolution Algorithms. – The Fast Fourier Transform. – Linear Filtering Computation of Discrete Fourier Transforms. – Polynomial Transforms. – Computation of Discrete Fourier
Transform by Polynomial Transforms. – Number Theoretic Transforms. – References. – Subject Index.

Digital Pattern Recognition

Editor: K. S. Fu
With contributions by T. M. Cover, E. Diday, K. S. Fu, A. Rosenfeld,
J.-C. Simon, T. J. Wagner, J. S. Weszka, J. J. Wolf

1976. 54 figures, 4 tables. XI, 206 pages
(Communication and Cybernetics, Volume 10)
ISBN 3-540-07511-9

Contents:
Introduction. – Topics in Statistical Pattern Recognition. – Clustering
Analysis. – Syntactic (Linguistic) Pattern Recognition. – Picture Re-
cognition. – Speech Recognition and Understanding. – Subject
Index.

J. D. Markel, A. H. Gray, Jr.

Linear Prediction of Speech

1976. 129 figures. XII, 288 pages
(Communication and Cybernetics, Volume 12)
ISBN 3-540-07563-1

Contents:
Introduction. – Formulations. – Solutions and Properties. – Acoustic
Tube Modeling. – Speech Synthesis Structures. – Spectral Analysis. –
Automatic Formant Trajectory Estimation. – Fundamental Fre-
quency Estimation. – Computational Considerations in Analysis. –
Vocoders. – Further Topics.

Springer-Verlag
Berlin
Heidelberg
New York

The Computer in Optical Research

Methods and Applications

Editor: B. R. Frieden

1980. 92 figures, 13 tables. Approx. 400 pages
(Topics in Applied Physics, Volume 41)
ISBN 3-540-10119-5

Contents:
B. R. Frieden: Introduction. – *R. Barakat:* The Calculation of Integrals Encountered in Optical Diffraction Theory. – *B. R. Frieden:* Computational Methods of Probability and Statistics. – *A. K. Rigler, J. R. Pegis:* Optimization Methods in Optics. – *L. Mertz:* Computers and Optical Astronomy. – *W. J. Dallas:* Computer-Generated Holograms.

Computer Processing of Electron Microscope Images

Editor: P. W. Hawkes

1980. 116 figures, 2 tables. XIV, 296 pages
(Topics in Current Physics, Volume 13)
ISBN 3-540-09622-1

Contents:
P. W. Hawkes: Image Processing Based on the Linear Theory of Image Formation. – *W. O. Saxton:* Recovery of Specimen Information for Strongly Scattering Objects. – *J. E. Mellema:* Computer Reconstruction of Regular Biological Objects. – *W. Hoppe, R. Hegerl:* Three-Dimensional Structure Determination by Electron Microscopy (Nonperiodic Specimens). – *J. Frank:* The Role of Correlation Techniques in Computer Image Processing. – *R. H. Wade:* Holographic Methods in Electron Microscopy. – *M. Isaacson, M. Utlaut, D. Kopf:* Analog Computer Processing of Scanning Transmission Electron Microscope Images.

Image Reconstruction from Projections

Implementation and Applications

Editor: G. T. Herman

1979. 120 figures, 10 tables. XII, 284 pages
(Topics in Applied Physics, Volume 32)
ISBN 3-540-09417-2

Contents:
G. T. Herman, R. M. Lewitt: Overview of Image Reconstruction from Projections. – *S. W. Rowland:* Computer Implementation of Image Reconstruction Formulas. – *R. N. Bracewell:* Image Reconstruction in Radio Astronomy. – *M. D. Altschuler:* Reconstruction of the Global-Scale Three-Dimensional Solar Corona. – *F. T. Budinger, G. T. Gullberg, R. H. Huesman:* Emission Computed Tomography. – *E. H. Wood, J. H. Kinsey, R. A. Robb, B. K. Gilbert, L. D. Harris, E. L. Ritman:* Applications of High Temporal Resolution Computerized Tomography to Physiology and Medicine.

Springer-Verlag
Berlin
Heidelberg
New York